METHODS IN MOLECULAR BIOLOGY

For further volumes:
http://www.springer.com/series/7651

For over 35 years, biological scientists have come to rely on the research protocols and methodologies in the critically acclaimed *Methods in Molecular Biology* series. The series was the first to introduce the step-by-step protocols approach that has become the standard in all biomedical protocol publishing. Each protocol is provided in readily-reproducible step-by-step fashion, opening with an introductory overview, a list of the materials and reagents needed to complete the experiment, and followed by a detailed procedure that is supported with a helpful notes section offering tips and tricks of the trade as well as troubleshooting advice. These hallmark features were introduced by series editor Dr. John Walker and constitute the key ingredient in each and every volume of the *Methods in Molecular Biology* series. Tested and trusted, comprehensive and reliable, all protocols from the series are indexed in PubMed.

RNA-RNA and RNA-DNA Interactions

Edited by

Michiaki Hamada

Faculty of Science and Engineering, Waseda University, Tokyo, Japan; National Institute of Advanced Industrial Science and Technology, Tokyo, Japan

Tsukasa Fukunaga

Waseda Institute for Advanced Study, Waseda University, Tokyo, Japan

Chao Zeng

Faculty of Science and Engineering, Waseda University, Tokyo, Japan

Humana Press

Editors
Michiaki Hamada
Faculty of Science and Engineering
Waseda University
Tokyo, Japan

National Institute of Advanced
Industrial Science and Technology
Tokyo, Japan

Tsukasa Fukunaga
Waseda Institute for Advanced Study
Waseda University
Tokyo, Japan

Chao Zeng
Faculty of Science and Engineering
Waseda University
Tokyo, Japan

ISSN 1064-3745 ISSN 1940-6029 (electronic)
Methods in Molecular Biology
ISBN 978-1-0716-4669-4 ISBN 978-1-0716-4670-0 (eBook)
https://doi.org/10.1007/978-1-0716-4670-0

This Humana imprint is published by the registered company Springer Science+Business Media, LLC, part of Springer Nature.
The registered company address is: 1 New York Plaza, New York, NY 10004, U.S.A.

Preface

Recent findings regarding noncoding RNAs have suggested that, much like proteins, these molecules appear to play a significant role in cellular processes. It has been observed that interactions with other biomolecules are crucial for the functionality of noncoding RNAs. While RNA-protein interactions have received the most attention in research, it is becoming increasingly evident that interactions between RNA and other nucleic acid molecules are also of importance. In this forthcoming volume, we endeavor to explore the dynamic and ever-evolving field of RNA-RNA and RNA-DNA interactions, which we believe are pivotal to our understanding of molecular biology and genetics. This book is structured into two sections—Experimental Approaches and Computational Approaches—each carefully curated to provide an in-depth exploration of the methodologies at the forefront of RNA biology.

Experimental Approaches offers an overview of the various innovative techniques that have been developed to elucidate the RNA-RNA and RNA-DNA interactions within the cellular context. Each chapter is detailed, providing thorough descriptions of the experimental designs and the interpretation of results. The aim here is to provide researchers with the practical knowledge they need to perform these techniques and to interpret complex data effectively. Computational Approaches, on the other hand, highlights the crucial role that computational tools and models play in complementing experimental data. This section showcases how computational biology can enhance our understanding of RNA-RNA and RNA-DNA interactions through predictive modeling and simulations. We have also included several chapters to provide details and illustrate the potential applications of these methodologies in biological research. By offering these advanced computational methods, we hope to facilitate a connection between theoretical knowledge and practical application, empowering researchers to utilize these tools to further their own studies.

The book presents these two approaches side by side, which may underscore the importance of a multidisciplinary approach in current scientific research. It also reflects the collaborative nature of contemporary research, where experimentalists and computational scientists work together to unravel the complexities of the biology. The integration of these methodologies may be crucial for driving forward our understanding of molecular mechanisms and for fostering innovations that may lead to new therapeutic strategies. This volume is intended to gently encourage researchers to explore and innovate in the field of RNA interactome, inspiring curiosity and motivating a deeper investigation into the myriad ways that RNA-RNA and RNA-DNA interactions shape biological systems.

In closing, this book provides the knowledge and tools necessary for researchers to explore the fascinating study of RNA-RNA and RNA-DNA interactions, aiming to inspire them to push beyond the boundaries of current scientific understanding. It hopes to serve as a valuable resource for researchers navigating the complex landscape of RNA interactome studies.

Tokyo, Japan — *Michiaki Hamada*
Tokyo, Japan — *Tsukasa Fukunaga*
Tokyo, Japan — *Chao Zeng*

Contents

Contributors

IÑAKI AMATRIA-BARRAL • *Computer Architecture Group, CITIC, Universidade da Coruña, Coruña, Spain*

ZHAOKUI CAI • *Key Laboratory of RNA Biology, Institute of Biophysics, Chinese Academy of Sciences, Beijing, China*

TSUKASA FUKUNAGA • *Graduate School of Advanced Science and Engineering, Waseda University, Tokyo, Japan*

ALEXEY A. GAVRILOV • *Institute of Gene Biology, Russian Academy of Sciences, Moscow, Russia*

JORGE GONZÁLEZ-DOMÍNGUEZ • *Computer Architecture Group, CITIC, Universidade da Coruña, Coruña, Spain*

MICHIAKI HAMADA • *Faculty of Science and Engineering, Waseda University, Shinjuku-ku, Tokyo, Japan; AIST-Waseda University Computational Bio Big-Data Open Innovation Laboratory (CBBD-OIL), Shinjuku-ku, Tokyo, Japan; Graduate School of Medicine, Nippon Medical School, Bunkyo-ku, Tokyo, Japan*

EITO ICHIHASHI • *Graduate School of Biological Sciences, Tokyo University of Science, Chiba, Japan*

YUKA W. IWASAKI • *Laboratory for Functional Non-coding Genomics, RIKEN Center for Integrative Medical Sciences, Yokohama, Japan*

YUKI KATO • *Department of RNA Biology and Neuroscience, Graduate School of Medicine, and Integrated Frontier Research for Medical Science Division (iFremed), Institute for Open and Transdisciplinary Research Initiatives (OTRI), The University of Osaka, Suita, Japan*

MAI KUBOTA • *Graduate School of Biological Sciences, Tokyo University of Science, Chiba, Japan*

WILSON H. LEE • *Department of Pharmacology and Pharmaceutical Sciences, Alfred E. Mann School of Pharmacy and Pharmaceutical Sciences, Los Angeles, CA, USA*

XINGWANG LI • *National Key Laboratory of Crop Genetic Improvement, Huazhong Agricultural University, Wuhan, Hubei, China*

JIE LIN • *Bioinformatics Section, School of Basic Medical Sciences, Southern Medical University, Guangzhou, China; College of Biological and Food Engineering, Guangdong University of Petrochemical Technology, Maoming, China*

ZHIPENG LU • *Department of Pharmacology and Pharmaceutical Sciences, Alfred E. Mann School of Pharmacy and Pharmaceutical Sciences, Los Angeles, CA, USA; Norris Comprehensive Cancer Center, University of Southern California, Los Angeles, CA, USA*

SERGEY MARGASYUK • *Center for Molecular and Cellular Biology, Skolkovo Institute of Science and Technology, Moscow, Russia*

KENSAKU MURANO • *Department of Molecular Biology, Keio University School of Medicine, Tokyo, Japan*

DMITRI D. PERVOUCHINE • *Center for Molecular and Cellular Biology, Skolkovo Institute of Science and Technology, Moscow, Russia*

SERGEY V. RAZIN • *Institute of Gene Biology, Russian Academy of Sciences, Moscow, Russia; Lomonosov Moscow State University, Moscow, Russia*

MASAYUKI SAKURAI • *Graduate School of Biological Sciences, Tokyo University of Science, Chiba, Japan*
KENGO SATO • *School of Life Science and Technology, Institute of Science Tokyo, Tokyo, Japan*
RICHARD A. SCHÄFER • *Feinberg School of Medicine, Northwestern University, Robert H. Lurie Medical Research Center, Chicago, IL, USA*
YUSUKE SHIROMOTO • *Department of Molecular Genetics, Graduate School of Medicine, Kyoto University, Kyoto, Japan*
TOMOKO TAKAHASHI • *Department of Biochemistry and Molecular Biology, Graduate School of Science and Engineering, Saitama University, Saitama, Japan*
AKITO TANEDA • *Graduate School of Science and Technology, Hirosaki University, Hirosaki, Aomori, Japan*
JUAN TOURIÑO • *Computer Architecture Group, CITIC, Universidade da Coruña, Coruña, Spain*
BJÖRN VOß • *Institute of Biomedical Genetics, University of Stuttgart, Stuttgart, Germany*
YUJIAN WEN • *Bioinformatics Section, School of Basic Medical Sciences, Southern Medical University, Guangzhou, China*
QIN XIAO • *National Key Laboratory of Crop Genetic Improvement, Huazhong Agricultural University, Wuhan, Hubei, China*
YUANCHAO XUE • *Key Laboratory of RNA Biology, Institute of Biophysics, Chinese Academy of Sciences, Beijing, China; University of Chinese Academy of Sciences, Beijing, China*
CHAO ZENG • *Faculty of Science and Engineering, Waseda University, Shinjuku-ku, Tokyo, Japan*
HAI ZHANG • *Network Center, Southern Medical University, Guangzhou, China*
MINJIE ZHANG • *Department of Bioinformatics, School of Basic Medical Sciences, Tianjin Medical University, Tianjin, China*
HAO ZHU • *Bioinformatics Section, School of Basic Medical Sciences, Southern Medical University, Guangzhou, China*

Part I

Experimental Approaches to Study RNA-RNA and RNA-DNA Interactions

Chapter 1

Protein-Centric Mapping of RNA–DNA Interactions with RedChIP

Alexey A. Gavrilov and Sergey V. Razin

Abstract

Noncoding RNAs (ncRNAs) are involved in a variety of processes in the cell nucleus, including transcriptional control, shaping 3D genome, and assembly/maintenance of functional nuclear compartments. However, the specific functions of most of the currently identified ncRNAs remain unclear. To gain further insight into the role of ncRNA in the eukaryotic genome functioning, it is important to determine genome-wide association patterns of various ncRNAs with chromatin. To address this question, a panel of RNA–DNA proximity ligation-based approaches have been developed, which have allowed deciphering RNA-chromatin interactions at the genome-wide level. An important drawback of all these techniques, however, is that they do not reveal the proteins involved in RNA–DNA interactions. In this chapter, we describe RedChIP, a method combining RNA–DNA ligation with chromatin immunoprecipitation to identify RNA–chromatin interactions mediated by a particular protein. We present a detailed protocol for RedChIP, focusing on the technical nuances and subtleties of the experimental procedure, and discuss algorithms for processing and analyzing the sequencing data. RedChIP can be used to identify RNAs associated with genomic regions occupied by any protein of interest, enabling disclosure of ncRNAs involved in recruiting various protein complexes to chromatin.

Key words Cell nucleus, Noncoding RNA, Chromatin, RNA–DNA interactome, Immunoprecipitation

1 Introduction

Recent studies have shown that the vast majority of the eukaryotic genome is transcribed to form a wide range of both coding and noncoding RNAs. Noncoding RNAs (ncRNAs) have been found to play a role in various biological processes in both the cytoplasm and the cell nucleus. Nuclear ncRNAs have been implicated in locus-specific gene activation and silencing, spatial organization of the genome, promoter–enhancer communication, organization of transcriptional hubs and heterochromatin, sustaining genome imprinting, maintaining genome stability, and the assembly of

Michiaki Hamada et al. (eds.), *RNA-RNA and RNA-DNA Interactions*, Methods in Molecular Biology, vol. 2949, https://doi.org/10.1007/978-1-0716-4670-0_1,

nuclear bodies [1–7]. Many of these processes seem to be mediated via the ability of ncRNAs to drive the process of liquid–liquid phase separation [8, 9].

A large amount of data on the functions of nuclear ncRNA has been obtained using methods for mapping the regions of association of ncRNAs with chromosomes (reviewed in Ref. [6]). The first bunch of these methods relied on cross-linking of RNA–protein–DNA complexes in vivo, followed by the pull-down of specific complexes that contain a particular RNA using biotinylated complementary oligonucleotides with subsequent identification of the associated DNA fragments via sequencing [10]. These methods allowed only one RNA to be examined in one experiment ("one-vs-all" approaches ChIRP [11], Chart [12], and RAP [13]). More recently, inspired by advances in 3D genomics achieved using HiC and other DNA–DNA proximity ligation methods [14, 15], several research groups have developed RNA–DNA proximity ligation approaches to simultaneously map genomic binding sites of all chromatin-bound RNA molecules, that is, RNA–DNA interactome ("all-vs-all" approaches: MARGI [16], GRID-seq [17], ChAR-seq [18], RADICL-seq [19], and Red-C [20]). Data collected from RNA–DNA proximity ligation experiments were used to investigate the cromosomal distribution of various ncRNAs, including MALAT1, RoX, XIST, enhancer RNAs, and spliceosomal RNAs, and to identify new ncRNAs involved in transcriptional regulation and genome spatial organization.

All the above-cited techniques, however, have a drawback in that they do not examine proteins involved in RNA–DNA interactions. Here, we describe a hybrid approach, RedChIP [21], which addresses this problem by combining an RNA–DNA proximity ligation protocol (Red-C [20]) with chromatin immunoprecipitation (ChIP), which allows the identification of ncRNAs associated with genomic regions occupied by specific proteins. Using RedChIP with antibodies against two human proteins, architectural protein CTCF and EZH2, a subunit of repressor complex PRC2, we uncovered ncRNAs interacting with DNA at the sites of deposition of these proteins, which, therefore, may be involved in chromatin looping and Polycomb-mediated gene silencing [21].

The RedChIP protocol is analogous to HiChIP and PLAC-seq used for mapping DNA–DNA interactions mediated by a particular protein [22, 23], with the difference that RNA–DNA interactions are analyzed instead. Briefly, RNA–protein–DNA complexes are fixed with formaldehyde in vivo, DNA is fragmented with a restriction enzyme, and the ends are blunted and dA-tailed (Fig. 1a). RNA 3′ ends are ligated to a bridge adapter containing a biotinylated nucleotide (Fig. 1b), followed by ligation of the opposite ends of the bridges with DNA ends in spatial proximity. Ligated complexes are solubilized by sonication and subjected to immunoprecipitation (IP) using antibodies against a protein of interest. RNA–

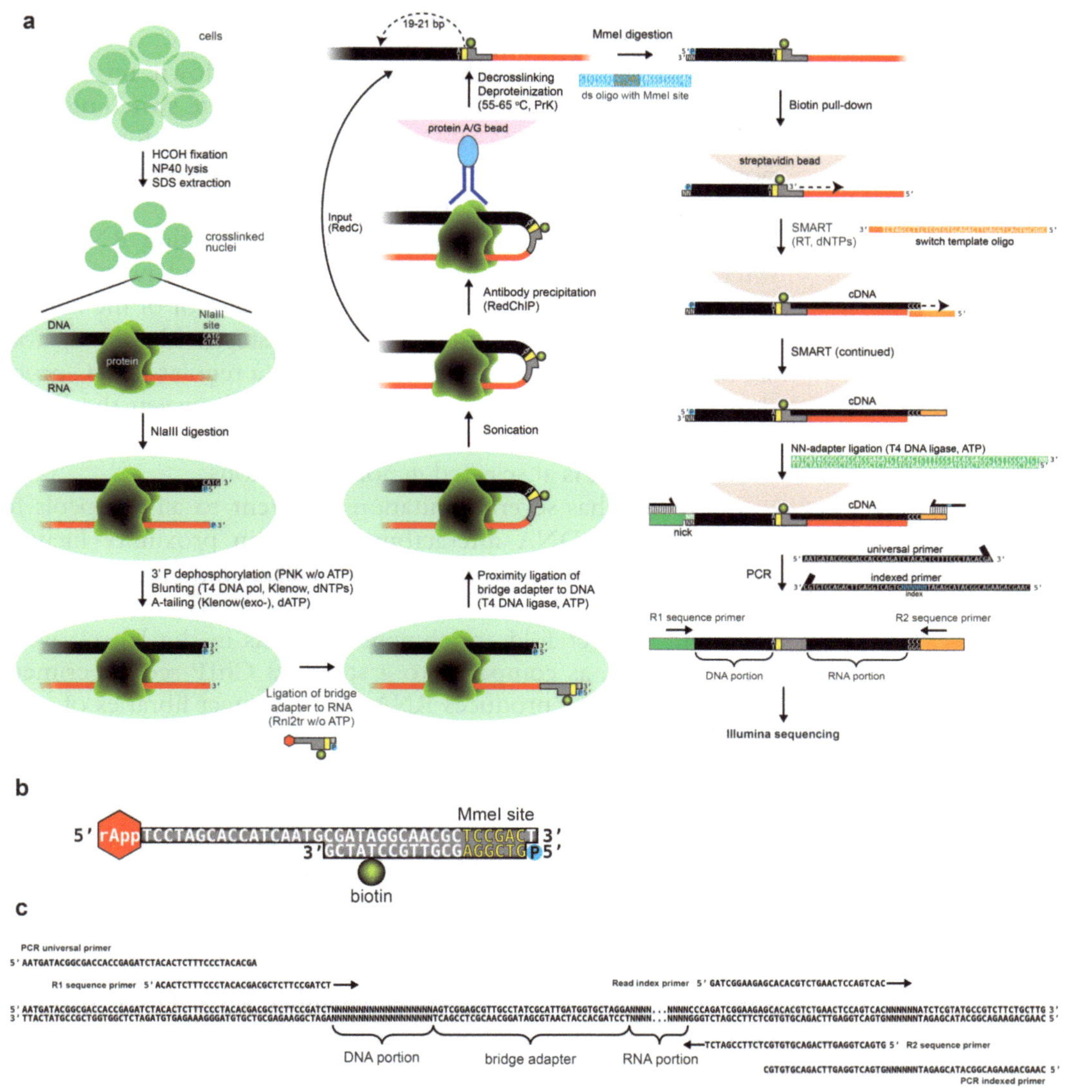

Fig. 1 Outline of the RedChIP protocol. (**a**) Main steps of the experimental procedure. (**b**) Bridge adapter structure. Bridge adapter is a duplex consisting of a 37-mer DNA oligonucleotide bearing an adenylate group (rApp) at the 5′ end and a hydroxyl group at the 3′ and an 18-mer DNA oligonucleotide bearing a phosphate group (P) at the 5′ end, a hydroxyl group at the 3′ end, and a biotin-modified internal thymidine residue. The single-stranded 5′-rApp end of the bridge adapter is used for ligation with RNA, while the double-stranded end of the bridge adapter with protruding T is used for the ligation with dA-tailed DNA fragments and also contains the MmeI recognition site (highlighted in yellow). The biotin is used for the pull-down of RNA–DNA chimeras, and the 3′ hydroxyl of the biotinylated strand is used to prime reverse transcription. (**c**) Sequencing library structure. Indicated are positions of DNA and RNA portions, bridge adapter, and PCR and sequencing primers

DNA chimeras are purified, and excess DNA is cut off using the MmeI restriction enzyme, the recognition site of which is incorporated into the bridge. After biotin pull-down, reverse transcription is primed from the bridge with template switching at the 5′ end of the RNA (SMART technology), allowing for the incorporation of a custom Illumina adapter (Fig. 1a). Finally, another Illumina adapter is ligated to the DNA ends, and the chimeras are amplified and paired-end sequenced, thus reporting RNA–DNA interactions mediated by a particular protein and protein–DNA interactions mediated by various RNAs. An aliquot of the material (input fraction) is processed without the IP step to identify the total set of RNA–DNA contacts in the cell, as in a regular Red-C procedure. By comparing the number of contacts between the IP and input fractions, it is then possible to calculate the fold enrichment of different RNAs at selected genomic locations.

RedChIP has several limitations inherent to all protocols for mapping RNA–DNA interactome based on proximity ligation. First, the high level of nonspecific background ligation in the procedure should be noted, which leads to a lower signal-to-noise ratio and a lower fold enrichment of RNAs in the IP relative to input fraction, as compared to a standard ChIP-seq experiment. Second, RedChIP produces RNA–DNA contact libraries of enormous complexity, meaning that very deep sequencing is required to map enough captured contacts for any specific, especially low abundant RNA. The problem is additionally complicated by the presence of multi-mapped RNA and DNA fragments that reduce the yield of unique RNA–DNA contacts. Finally, RedChIP relies on chemical cross-linking, which can potentially introduce artifacts [24].

It should also be noted that some of the information obtained with RedChIP can be gained in a regular RNA–DNA proximity ligation experiment accompanied by ChIP-seq, with subsequent comparison of the genomic distribution profiles of the RNA and protein. Still, RedChIP may be advantageous in some cases, for example, when searching for rare RNAs associated with a rare chromatin protein, which can be undiscoverable in the total RNA–DNA interactome, or when multiple protein-binding sites overlap on DNA making it impossible to tell which protein an RNA interacts with. The latter question can be addressed with RNA immunoprecipitation (RIP-seq [25]), but then again not quite accurately, as RIP-seq does not distinguish between situations when RNA is bound to a free or DNA-bound protein. Being in fact 3-in-1 (ChIP-seq, RIP-seq, and RNA–DNA profiling in a single experiment), RedChIP provides rich information on the interrelationship between RNA, DNA, and chromatin-bound proteins, making it a useful tool for exploring the functions of nuclear ncRNAs, especially those involved in targeting various proteins to chromatin.

2 Materials

2.1 Oligonucleotide Dissolving and Annealing

1. Oligonucleotides (purchased from IDT in dry form, Table 1).
2. Low-EDTA TE buffer: 10 mM Tris–HCl pH 8.0, 0.1 mM EDTA.
3. 5 M NaCl.

2.2 Cell Fixation and Lysis

1. 37% formaldehyde (Sigma-Aldrich, Cat. No. F8775).
2. 2.5 M glycine.
3. PBS.
4. Permeabilization buffer: 10 mM Tris–HCl pH 7.5, 10 mM NaCl, 0.2% NP40. Store at 4 °C.
5. 100× Protease inhibitor cocktail.
6. SUPERase·In™ RNase inhibitor (20 U/μL, Thermo Fisher Scientific).
7. 10× NEBuffer 4.
8. 10% SDS (w/v).
9. 20% Triton X-100 (v/v).
10. Nuclease-free water.

2.3 In Situ Restriction Enzyme Digestion and RNA–DNA Ligation

1. NlaIII (10 U/μL, NEB).
2. T4 PNK (10 U/μL, NEB).
3. T4 DNA Polymerase (3 U/μL, NEB).
4. Klenow (5 U/μL, NEB).
5. Klenow (exo-) (50 U/μL, NEB).
6. T4 Rnl2tr (200 U/μL, NEB).
7. T4 DNA Ligase (5 Weiss U/μL, Thermo Fisher Scientific).
8. 100 μM bridge adapter.
9. 10 mM dNTP mix (10 mM each of dATP, dCTP, dGTP, and dTTP).
10. 10 mM dATP.
11. 10× NEBuffer 4.
12. 10× NEBuffer 2.
13. 10× T4 PNK Buffer (NEB).
14. 10× T4 DNA Ligase Buffer (NEB).
15. 10× RNA Ligase Buffer (NEB).
16. 10× T4 DNA Ligase Buffer (Thermo Fisher Scientific).
17. 50% PEG-8000 (NEB).

Table 1
Sequences of oligos and adapters used in the RedChIP procedure

Name	Step	5′-3′ Sequence
ds bridge adapter	3.8	/rApp/TCCTAGCACCATCAATGCGATAGGCAACGCTCCGACT [a] /Phos/GTCGGAGCGTTGCC/T-Biotin/ATCG
ds oligo with MmeI site	3.13	CTGTCCGTTCCGACTACCCTCCCGAC GTCGGGAGGGTAGTCGGAACGGACAG
Switch template oligo	3.15	iCiGiCGTGACTGGAGTTCAGACGTGTGCTCTTCCGATCTrGrGrG[b]
ds NN-adapter	3.16	AGATCGGAAGAGCGTCGTGTAGGGAAAGAGTGTAGATCTCGGTGGTCGCCGTATCATT AATGATACGGCGACCACCGAGATCTACACTCTTTCCCTACACGACGCTCTTCCGATCTNN[c]
PCR universal primer	3.17	AATGATACGGCGACCACCGAGATCTACACTCTTTCCCTACACGA
PCR indexed primer	3.17	CAAGCAGAAGACGGCATACGAGATNNNNNNGTGACTGGAGTTCAGACGTGTGC[d]

[a]3′ hydroxyl non-blocked (*see* **Note 1**)
[b]iC and iG designate Iso-dC and Iso-dG, and r indicates ribonucleotides
[c]N designates any base
[d]NNNNNN is an Illumina TruSeq single index of choice, for example: index 7 GATCTG, index 8 TCAAGT, etc.

18. 10 mg/mL BSA (NEB).
19. 10% SDS (w/v).
20. 20% Triton X-100 (v/v).
21. Nuclease-free water.

2.4 Sonication and Immunoprecipitation

1. RIPA buffer: 50 mM Tris–HCl pH 7.5, 150 mM NaCl, 2 mM EDTA, 0.5% SDC, 0.1% SDS, 1% NP-40. Store at 4 °C.
2. ChIP-Grade antibody against a protein of interest.
3. Pierce ChIP-Grade Protein A/G Magnetic Beads (10 mg/mL, Thermo Fisher Scientific).
4. 5% BSA (w/v) in nuclease-free water (for block solution). Store at 4 °C.
5. Wash Buffer A (0.1% SDS, 1% Triton X-100, 2 mM EDTA, 20 mM Tris–HCl pH 8.0, 150 mM NaCl). Store at 4 °C.
6. Wash Buffer B (0.1% SDS, 1% Triton X-100, 2 mM EDTA, 20 mM Tris–HCl pH 8.0, 500 mM NaCl). Store at 4 °C.
7. 100× Protease inhibitor cocktail.
8. SUPERase· In RNase inhibitor (20 U/μL, Thermo Fisher Scientific).
9. Amicon 30 K Ultra-0.5 mL Centrifugal Filters (Millipore).

2.5 RNA–DNA Chimera Purification and MmeI Digestion

1. Proteinase K (20 mg/mL, Ambion).
2. MmeI (2 U/μL, NEB).
3. 5 M NaCl.
4. 1 M Tris–HCl pH 7.5.
5. 0.5 M EDTA.
6. 10% (w/v) SDS.
7. GlycoBlue (15 mg/mL, Thermo Fisher Scientific).
8. 3 M NaAc.
9. Isopropanol.
10. AMPure XP beads (Beckman Coulter).
11. 80% Ethanol.
12. Qubit dsDNA High Sensitivity Assay Kit.
13. 10× NEBuffer 4.
14. 10 mg/mL BSA (NEB).
15. 32 μM SAM (NEB).
16. 50 μM ds oligo with MmeI site.
17. Nuclease-free water.

2.6 Biotin Pull-Down and Library Construction

1. Dynabeads MyOne Streptavidin C1 beads (10 mg/mL, Thermo Fisher Scientific).
2. SMARTScribe Reverse Transcriptase (100 U/μL, Clontech).
3. T4 DNA Ligase (5 Weiss U/μL, Thermo Fisher Scientific).
4. KAPA HiFi DNA Polymerase (1 U/μL, Roche).
5. Tween Washing Buffer (TWB): 5 mM Tris–HCl pH 7.5, 0.5 mM EDTA, 1 M NaCl, 0.05% Tween 20.
6. 2× Binding Buffer: 10 mM Tris–HCl pH 7.5, 1 mM EDTA, 2 M NaCl.
7. 10× NEBuffer 2.
8. 5× First-Strand Buffer (Clontech).
9. 20 mM DTT (Clontech).
10. 10 mM dNTP mix (10 mM each of dATP, dCTP, dGTP, and dTTP).
11. 100 μM switch template oligo.
12. SUPERase· In RNase inhibitor (20 U/μL, Thermo Fisher Scientific).
13. 10× T4 DNA Ligase Buffer (Thermo Fisher Scientific).
14. 5× Rapid Ligation Buffer from Rapid Ligation Kit (Thermo Fisher Scientific).
15. 50 μM NN-adapter.
16. 5× KAPA HiFi Fidelity Buffer (Roche).
17. 20 μM PCR universal primer.
18. 20 μM PCR indexed primer.
19. AMPure XP beads (Beckman Coulter).
20. 80% Ethanol.
21. 10 mM Tris–HCl pH 8.0.
22. Nuclease-free water.

3 Methods

All work is carried out in an RNase-free environment using RNase-free reagents and materials.

3.1 Oligonucleotide Dissolving and Annealing

Dissolve oligos in low-EDTA TE buffer at 200 μM (bridge adapter forward and reverse oligos) or 100 μM (all the other oligos). For PCR, prepare 20 μM working solutions of primers by diluting 100 μM stock 1:5 with low EDTA TE buffer.

To prepare a duplex, use the following protocol:

1. Mix 50 μL of a forward oligo, 50 μL of a reverse oligo, and 1 μL of 5 M NaCl.
2. Divide the mixture between two PCR tubes and incubate in a PCR machine with a hot lid using the following program:

Step	Temperature	Time	Cycles
Denaturation	94 °C	2 min	1
Annealing	94 °C (−1 °C/cycle) 4 °C	30 s Pause	70 1

Store all oligos at −20 °C.

3.2 Cell Fixation

Use 4–6 × 10^7 cells per experiment (*see* **Note 2**).

3.2.1 Suspension Cells

1. Prepare a single-cell suspension in 20 mL of full growth media.
2. Add an equal volume of freshly prepared 2× fix solution (2% formaldehyde in full-growth media) and incubate for 10 min at RT (here and elsewhere 22 °C) with rotation.
3. Stop fixation by adding 2.5 M glycine to a concentration of 0.125 M and incubate for 5 min at RT with rotation, then for 5 min on ice.
4. Pellet the cross-linked cells by centrifugation for 10 min at 300× *g* and 4 °C.
5. Wash the cells by resuspending in 40 mL of cold PBS and repeat centrifugation (*see* **Note 3**).

3.2.2 Adherent Cells

1. Remove the growth medium from the cell monolayer.
2. Add 10 mL of freshly prepared 1× fix solution (1% formaldehyde in full-growth media or PBS) per 10 cm culture dish or 75 cm^2 culture flask and incubate for 10 min at RT with agitation.
3. Stop fixation by adding 2.5 M glycine to a concentration of 0.125 M and incubate for 5 min at RT with agitation.
4. Wash the cross-linked cells with 10 mL of cold PBS.
5. Scrap the cells with a cell scraper into 10 mL of cold PBS and transfer the suspension to a 15-mL Falcon tube.
6. Pellet the cells by centrifugation for 10 min at 300× *g* and 4 °C (*see* **Note 3**).

3.3 Cell Lysis

1. Resuspend the cell pellet in 4 mL of cold permeabilization buffer supplemented, prior to use, with 1× protease inhibitors and 400 U of SUPERase· In RNase inhibitor, and incubate for 10 min on ice.

2. Pellet the permeabilized cells by centrifugation for 3 min at 2500× *g* and 4 °C in a cellular centrifuge.
3. Resuspend the cell pellet in 900 μL of nuclease-free water by pipetting up and down and transfer the sample to a 1.5-mL Eppendorf tube.
4. Add 27 μL of 10% SDS and incubate for 30 min at 37 °C with shaking at 1200 rpm in a temperature-controlled shaker (*see* **Note 4**).
5. Add 90 μL of 20% Triton X-100 and incubate for 30 min at 37 °C with shaking at 1200 rpm to sequester the SDS.
6. Add 360 μL of warm 4× NEBuffer 4 and incubate for 3 min at 37 °C with shaking at 1200 rpm (*see* **Note 5**).
7. Pellet the nuclei by centrifugation for 4 min at 2500× *g* and RT.

3.4 Restriction Enzyme Digestion

1. Resuspend the nuclear pellet in 1250 μL of 1× NEBuffer 4.
2. Take a 50-μL aliquot for the DNA integrity control (optional) and divide the remaining mixture into two 600-μL aliquots (*see* **Note 6**).
3. Add 30 μL of NlaIII restriction enzyme (10 U/μL, NEB) to each 600-μL aliquot and incubate, along with the DNA integrity control, for 3–4 h at 37 °C with shaking at 1200 rpm (*see* **Note 7**).
4. Pool the 600-μL aliquots and take 50 μL for the DNA digestion control (optional).
5. Centrifuge the main and control samples for 4 min at 2500× *g* and RT and remove the supernatant.
6. (Optional) Isolate DNA from the control samples and check the integrity of the non-digested DNA control and the efficacy of the restriction enzyme digestion using an agarose gel electrophoresis (*see* **Note 8**).
7. Resuspend the nuclear pellet in 600 μL of 1× NEBuffer 2.
8. Add 15 μL of 10% SDS and mix thoroughly by pipetting up and down to inactivate residual restriction enzyme.
9. Add 35 μL of 20% Triton X-100 and incubate for 5 min at 37 ° C with shaking at 1200 rpm.
10. Pellet the nuclei by centrifugation for 4 min at 2500 × *g* and RT.
11. Wash the nuclei by resuspending in 600 μL of 1× T4 PNK Buffer (NEB) supplemented with 0.1% Triton X-100, transfer the mixture to a new 1.5-mL Eppendorf tube, and pellet as above.

3.5 RNA 3′ P Dephosphorylation

Refer to **Note 9** for the purpose of the dephosphorylation step.

1. Resuspend the nuclear pellet in 380 μL of the dephosphorylation solution prepared as follows:

Component	For one reaction (μL)
10× T4 PNK buffer (NEB)	40
20% triton X-100	2
Nuclease-free water	338
Total volume	380

2. Add 20 μL of T4 PNK (10 U/μL, NEB), mix by pipetting, and incubate for 30 min at 37 °C with shaking at 900 rpm.
3. Pellet the nuclei by centrifugation for 4 min at 2500× *g* and RT.

3.6 DNA End Blunting

1. Resuspend the nuclear pellet in 378 μL of the following blunting solution:

Component	For one reaction (μL)
10× T4 DNA ligase buffer (NEB)	40
10 mM dNTP mix	10
Nuclease-free water	328
Total volume	378

2. Add 10 μL of T4 DNA Polymerase (3 U/μL, NEB) and 12 μL of Klenow (5 U/μL, NEB), mix by pipetting, and incubate for 1 h at RT with shaking at 900 rpm (*see* **Note 10**).
3. Add 10 μL of 10% SDS, mix by pipetting, and centrifuge for 4 min at 2500× *g* and RT.
4. Wash the nuclei with 600 μL of 1× NEBuffer 2 supplemented with 1% Triton X-100, transfer the mixture to a new 1.5-mL Eppendorf tube, and pellet as above.

3.7 DNA End dA Tailing

1. Resuspend the nuclear pellet in 397 μL of the following dA-tailing solution:

Component	For one reaction (μL)
10× NEBuffer 2	40
10 mM dATP	20
20% triton X-100	20
Nuclease-free water	317
Total volume	397

2. Add 3-μL Klenow (exo-) (50 U/μL, NEB), mix by pipetting, and incubate for 1 h at 37 °C with shaking at 900 rpm.
3. Add 10 μL of 10% SDS, mix by pipetting, and centrifuge for 4 min at 2500× *g* and RT.
4. Wash the nuclei with 600 μL of 1× NEBuffer 2 supplemented with 1% Triton X-100 and pellet as above.
5. Wash the nuclei with 600 μL of 1× RNA Ligase Buffer (NEB) supplemented with 0.1% Triton X-100 and pellet as above.
6. Wash the nuclei with 600 μL of 1× RNA Ligase Buffer (NEB), transfer the mixture to a new 1.5-mL Eppendorf tube, and pellet as above (*see* **Note 11**).

3.8 RNA-to-Bridge Adapter Ligation

1. Resuspend the nuclear pellet in 376 μL of the following RNA Ligase solution (*see* **Note 12**):

Component	For one reaction (μL)
10× RNA ligase buffer (NEB)	40
100 μM bridge adapter	18
50% PEG-8000 (NEB)	160
Nuclease-free water	158
Total volume	376

2. Add 24 μL of T4 Rnl2tr (200 U/μL, NEB), mix by pipetting, and incubate for 6 h at RT then overnight at 16 °C with shaking at 900 rpm (*see* **Note 13**).
3. Pellet the nuclei by centrifugation for 4 min at 2500× *g* and RT.
4. Resuspend the nuclear pellet in 600 μL of 0.1% SDS supplemented with 0.1 mg/mL BSA (NEB), incubate for 4 min at 37 °C with shaking at 1200 rpm, and pellet as above (*see* **Note 14**).
5. Repeat the previous step two more times. Add 30 μL of 20% Triton X-100 before the final centrifugation.
6. Further, wash the nuclei twice with 600 μL of 1× T4 DNA Ligase Buffer (Thermo Fisher Scientific) supplemented with 0.1% Triton X-100 by repeating resuspending/pelleting. Transfer the mixture to a new 1.5-mL Eppendorf tube before the final centrifugation (*see* **Note 15**).

3.9 Bridge Adapter-to-DNA Ligation (Proximity Ligation)

1. Resuspended the nuclear pellet in 285 μL of the following phosphorylation solution:

Component	For one reaction (μL)
10× T4 DNA ligase buffer (NEB)	30
20% triton X-100	1.5

(continued)

Component	For one reaction (μL)
Nuclease-free water	253.5
Total volume	285

2. Add 15 μL of T4 PNK (10 U/μL, NEB), mix by pipetting, and incubate for 1 h at 37 °C with shaking at 900 rpm (*see* **Note 16**).
3. Pellet the nuclei by centrifugation for 4 min at 2500× *g* and RT.
4. Resuspend the nuclear pellet in 1 mL of 1× T4 DNA Ligase Buffer (Thermo Fisher Scientific).
5. Add 20 μL of T4 DNA Ligase (5 Weiss U/μL, Thermo Fisher Scientific), mix by pipetting, and incubate for 6 h at RT with rotation.
6. Pellet the nuclei as above, wash with 400 μL of 1× T4 DNA Ligase Buffer (Thermo Fisher Scientific), and pellet again.

3.10 Sonication

1. Resuspend the nuclear pellet in 1 mL of RIPA buffer supplemented, prior to use, with 1× protease inhibitors and 100 U of SUPERase· In RNase inhibitor.
2. Sonicate intensively with an appropriate sonication device keeping the sample cool (*see* **Note 17**).
3. Centrifuge at 16,100× *g* for 5 min at 4 °C to remove non-solubilized material.

3.11 Immunoprecipitation

1. Transfer the supernatant to two Amicon 30 K Ultra-0.5 mL Centrifugal Filters (500 μL per filter) and centrifuge at 16,100× *g* for 5 min at 4 °C.
2. Discard the flow-through, add 400 μL of RIPA buffer to each filter, and repeat centrifugation.
3. Collect the concentrated solution from the two filters, combine, and set aside 1/20 as an input fraction (store at 4 °C).
4. Split the remaining solution evenly between four 1.5-mL Eppendorf tubes and add RIPA buffer to a final volume of 1 mL in each tube.
5. Add 4 μL of antibody to each tube and incubate overnight at 4 °C with rotation.
6. While the previous step is in progress, mix 160 μL of Pierce ChIP-Grade Protein A/G Magnetic Beads with 1 mL of freshly prepared block solution (0.9× RIPA, 0.5% (w/v) BSA) and incubate for 10 min at 4 °C with rotation.
7. Separate the beads on a magnet and discard the supernatant.

8. Add 1 mL of block solution, vortex to resuspend, and incubate for 10 min at 4 °C with rotation. Separate the beads on a magnet and discard the supernatant.
9. Add 1 mL of block solution, vortex to resuspend, and incubate overnight at 4 °C with rotation.
10. Split the bead suspension evenly between four 1.5-mL Eppendorf tubes, separate the beads on a magnet, and discard the supernatant.
11. Add to the beads the IP reactions from **step 5** of this section, vortex to resuspend, and incubate for 6 h at 4 °C with rotation. Separate the beads on a magnet and discard the supernatant.
12. To wash the beads, add 1 mL of cold RIPA buffer, resuspend the beads by vortexing, and incubate for 10 min at 4 °C with rotation. Separate the beads on a magnet and discard the supernatant.
13. Wash the beads with 1 mL of cold Wash Buffer A using the same procedure as in the previous step.
14. Wash the beads with 1 mL of cold Wash Buffer B as above.

3.12 RNA–DNA Chimera Purification

1. Resuspend the beads in 117.5 μL of Proteinase K solution prepared according to the recipe below and combine the four samples into one tube (IP sample).

Component	For one reaction (μL)
5 M NaCl	10
1 M Tris–HCl pH 7.5	5
0.5 M EDTA	2
10% SDS	50
Nuclease-free water	403
Total volume	470

2. Add Proteinase K solution (prepared as above) to the input sample from **step 3** of Subheading 3.11 to a final volume of 470 μL.
3. Add 30-μL Proteinase K (20 mg/mL, Ambion) to the IP and input samples, mix by pipetting, and incubate for 1 h at 55 °C and then for 2 h at 65 °C with shaking at 900 rpm. During incubation, spin down briefly every 30–60 min to collect any condensate from the tube caps.
4. Cool to RT, place the IP sample on a magnet, and transfer the clear solution to a new 1.5-mL Eppendorf tube.

5. Add 3 μL of GlycoBlue, 50 μL of 3 M NaAc, and 550 μL of isopropanol to the IP and input samples. Mix by inverting and incubate overnight at −80 °C.
6. Centrifuge for 30 min at 21,000× *g* and 4 °C and remove the supernatant (*see* **Note 18**).
7. Resuspend the pellet in 50 μL of nuclease-free water.
8. Add 2 volumes of AMPure XP beads, mix by pipetting, and incubate for 5 min at RT.
9. Separate the beads on a magnet and discard the supernatant.
10. Add 0.7 mL of freshly prepared 80% ethanol, remove the tube from the magnet, and invert several times. Place the tube back in the magnet and discard the supernatant.
11. Repeat wash with 80% ethanol one more time.
12. Spin down the tube at 500× *g* for 10 s, place in a magnet, and remove the residual ethanol.
13. Keeping the tube on the magnet, air dry the beads for 10 min at RT.
14. Resuspend the beads in 50 μL of nuclease-free water by pipetting up and down and incubate for 5 min at RT.
15. Separate the beads on a magnet and transfer the eluent to a new Eppendorf low-binding tube.
16. Quantify concentration by Qubit dsDNA High Sensitivity Assay.

3.13 MmeI Digestion

1. Set up two 60-μL MmeI digestion reactions (one for the IP sample and the other for the input sample) as specified below (*see* **Note 19**). Use all of the IP sample and 500–1000 ng of the input sample per reaction. To avoid pipetting too small volumes of SAM and ds oligo stock solutions, prepare their fresh 1:10 dilutions by mixing 1 μL of the stock with 9 μL of nuclease-free water.

Component	For one reaction (μL)
RNA–DNA chimeras	Variable
10× NEBuffer 4	6
10 mg/mL BSA (NEB)	0.6
32 μM SAM (NEB), 1:10 dilution	1.5
50 μM ds oligo with MmeI site, 1:10 dilution	1.2
MmeI (2 U/μL, NEB)	1
Nuclease-free water	Variable
Total volume	60

2. Incubate for 2 h at 37 °C.

3.14 Biotin Pull-Down

All the following steps are performed in 1.5-mL Eppendorf low-binding tubes using low-binding pipet tips.

1. Wash 12 μL of Dynabeads MyOne Streptavidin C1 beads (for two samples) with 400 μL of Tween Washing Buffer (TWB) by pipetting up and down and incubating for 3 min at RT.
2. Separate the beads on a magnet and discard the supernatant.
3. Resuspend the beads in 400 μL of TWB by pipetting and transfer the mixture to a new tube. Separate the beads on a magnet and discard the supernatant.
4. Resuspend the beads in 125 μL of 2× Binding Buffer (for two reactions plus an extra 5 μL).
5. Add 60 μL of the beads to MmeI digestion reactions from **step 2** of Subheading 3.13 and incubate for 15 min at RT, vortexing gently every 5 min. Separate the beads on a magnet and discard the supernatant.
6. Wash the beads with 600 μL of TWB and transfer the mixture to a new tube. Separate the beads on a magnet and discard the supernatant.
7. Repeat wash with TWB.
8. Wash the beads with 100 μL of 1× NEBuffer 2 and transfer the mixture to a new tube. Separate the beads on a magnet and discard the supernatant.
9. Wash the beads with 50 μL of 1× First-Strand Buffer (Clontech) and transfer the mixture to a new tube. Separate the beads on a magnet and discard the supernatant.

3.15 Reverse Transcription (SMART)

1. Resuspend the beads with tethered RNA–DNA chimeras in 38 μL of the following reverse transcriptase solution:

Component	For one reaction (μL)	Master mix for 2.2 reactions (μL)
5× first-Strand buffer (Clontech)	8	17.6
20 mM DTT (Clontech)	5	11
10 mM dNTP mix	4	8.8
100 μM switch template oligo	0.4	0.88
SUPERase· In RNase inhibitor	1	2.2
Nuclease-free water	19.6	43.1
Total volume	38	38/tube

2. Preheat the mixture for 2 min at 42 °C.
3. Add 2 μL of SMARTScribe Reverse Transcriptase (100 U/μL, Clontech), mix by pipetting, and incubate for 1 h at 42 °C with shaking at 800 rpm (*see* **Note 20**). Vortex the mixture gently in the middle and in the end of the incubation.
4. Separate the beads on a magnet and discard the supernatant.
5. Wash the beads with 600 μL of TWB and transfer the mixture to a new tube. Separate the beads on a magnet and discard the supernatant.
6. Repeat wash with TWB.
7. Wash the beads with 100 μL of 1× NEBuffer 2 and transfer the mixture to a new tube. Separate the beads on a magnet and discard the supernatant.
8. Wash the beads with 100 μL of 1× T4 DNA Ligase Buffer (Thermo Fisher Scientific) and transfer the mixture to a new tube. Separate the beads on a magnet and discard the supernatant.

3.16 Illumina Adapter Ligation

1. Resuspend the beads in 48 μL of the following DNA Ligase solution:

Component	For one reaction (μL)	Master mix for 2.2 reactions (μL)
5× rapid ligation buffer (Thermo fisher scientific)	10	22
50 μM NN adapter	3	6.6
Nuclease-free water	35	77
Total volume	48	48/tube

2. Add 2 μL of T4 DNA Ligase (5 Weiss U/μL, Thermo Fisher Scientific), mix by pipetting, and incubate for 30 min at RT (*see* **Note 21**). Vortex the mixture gently at the end of the incubation.
3. Separate the beads on a magnet and discard the supernatant.
4. Wash the beads with 600 μL of TWB and transfer the mixture to a new tube. Separate the beads on a magnet and discard the supernatant.
5. Repeat wash with TWB.
6. Wash the beads with 100 μL of 1× NEBuffer 2 and transfer the mixture to a new tube. Separate the beads on a magnet and discard the supernatant.

7. Wash the beads with 100 μL 10 mM Tris–HCl pH 8.0 and transfer the mixture to a new tube. Separate the beads on a magnet and discard the supernatant.
8. Finally, resuspend the beads in 12 μL of water and store at – 20 °C.

3.17 Library Amplification and Sequencing

1. Set up a 50-μL analytical PCR reaction by mixing 4 μL of the beads from **step 8** of Subheading 3.16 with 46 μL of the following PCR solution (*see* **Note 22**):

Component	For one reaction (μL)	Master mix for 2.2 reactions (μL)
5× KAPA HiFi Fidelity buffer	10	22
10 mM dNTP mix	1.5	3.3
20 μM PCR universal primer	1.25	2.75
20 μM PCR indexed primer	1.25	2.75
KAPA HiFi DNA polymerase (1 U/μL)	1	2.2
Nuclease-free water	31	68.2
Total volume	46	46/tube

2. Run PCR using the following program:

Step	Temperature (°C)	Time	Cycles
Initial denaturation	95	5 min	1
Denaturation Annealing Extension	98 65 72	20 s 15 s 20 s	10–22 cycles
Final extension	72	3 min	1
	4	Pause	1

Take 6-μL aliquots after 10, 13, 16, 19, and 22 cycles at the end of the extension step, pausing the PCR machine if necessary.

3. Run the collected aliquots on a 2% agarose gel along with a molecular weight DNA ladder. The results of a typical titration PCR for the IP and input samples are presented in Fig. 2.
4. Choose an optimal number of PCR cycles for the final amplification of the libraries. We recommend using a minimal number of PCR cycles when PCR products only begin to be visible on a gel (15 cycles for Fig. 2).

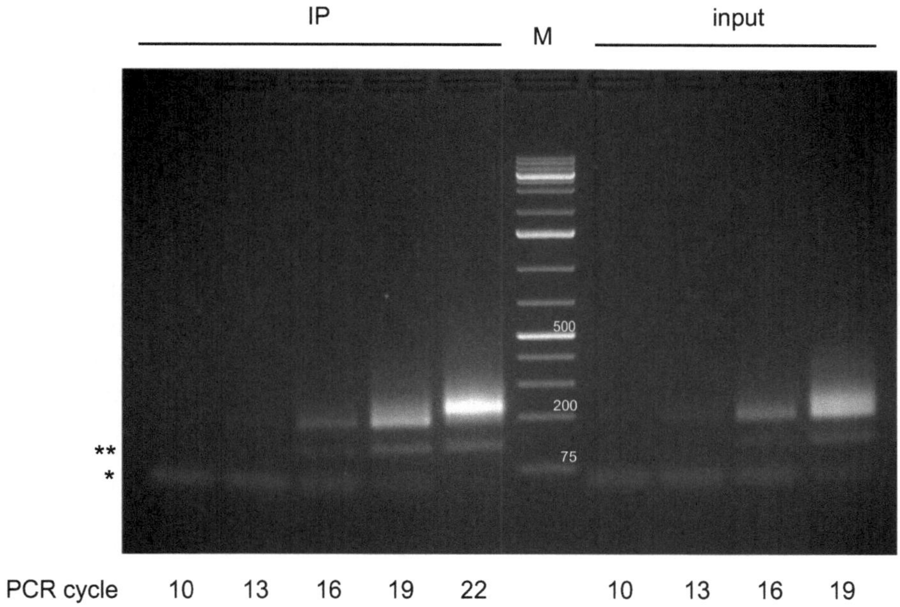

Fig. 2 Agarose gel showing typical results of titration PCR for IP and input samples. The gel contains products from amplification of RNA–DNA chimeras obtained after different number of PCR cycles (10, 13, 16, 19, and 22 for IP sample and 10, 13, 16, and 19 for input sample). RNA–DNA chimeras are expected to have a DNA portion of 18–20 nucleotides and an RNA portion of variable size (*see* Fig. 1c), yielding PCR products of ≥181 bp. An asterisk shows free primer, a double asterisk shows primer dimer; M, molecular weight marker (Thermo Fisher Scientific, Cat. No. SM1331). A total of 15 cycles were used for the final amplification of the libraries in this experiment

5. For the final amplification, set up two 50-μL PCR reactions as above for each sample, using a unique indexed primer for each sample.
6. Run PCR for the chosen number of cycles using the above program.
7. Pool the two PCR reactions, separate the streptavidin beads on a magnet, and transfer the clear solution containing amplification products to a new tube.
8. Add 1 volume of AMPure XP beads (Beckman Coulter), mix by pipetting, and incubate for 5 min at RT.
9. Separate the beads on a magnet and discard the supernatant.
10. Keeping the tube on the magnet, wash twice with 0.7 mL of freshly prepared 80% ethanol.
11. Spin down the tube at 500× *g* for 10 s, place the tube in a magnet, and remove the residual ethanol.
12. Keeping the tube on the magnet, air dry the beads at RT until the first signs of cracking appear.
13. Resuspend the beads in 50 μL of 10 mM Tris–HCl pH 8.0 and incubate for 5 min at RT.

14. Separate the beads on a magnet and transfer the eluent to a new tube.
15. Clean up the PCR products once more by repeating **steps 8–14** (*see* **Note 23**). Use 10 μL of 10 mM Tris–HCl pH 8.0 for the final DNA elution. A typical DNA yield is 50–100 ng.
16. Sequence the library with 2 × 100-nt paired-end reads on an Illumina platform (Fig. 1c).

3.18 Sequencing Data Processing

The sequencing data are processed using the RedClib software package (https://github.com/agalitsyna/RedClib) which converts raw sequencing reads into unique RNA-DNA contacts (*see* **Note 24**). The main steps of the RedClib pipeline are summarized in Fig. 3 and discussed below.

1. Filtering out PCR duplicates. Both forward (R1) and reverse (R2) reads are cut to the first 50 nucleotides. Then, Fast Uniq is used to search for exact duplicates. From a group of duplicates, only one pair is retained.
2. Quality trimming. We use TRIMMOMATIC for the detection of the first leftmost low-quality position in each forward and reverse read. The parameters are set to: window size: 5, quality threshold: 20. Only reads with at least one nucleotide passing the quality control filter are selected.
3. Scanning of reads for bridge, switch template oligo, and CCC/GGG trinucleotides. The scanning is done with Rabin–

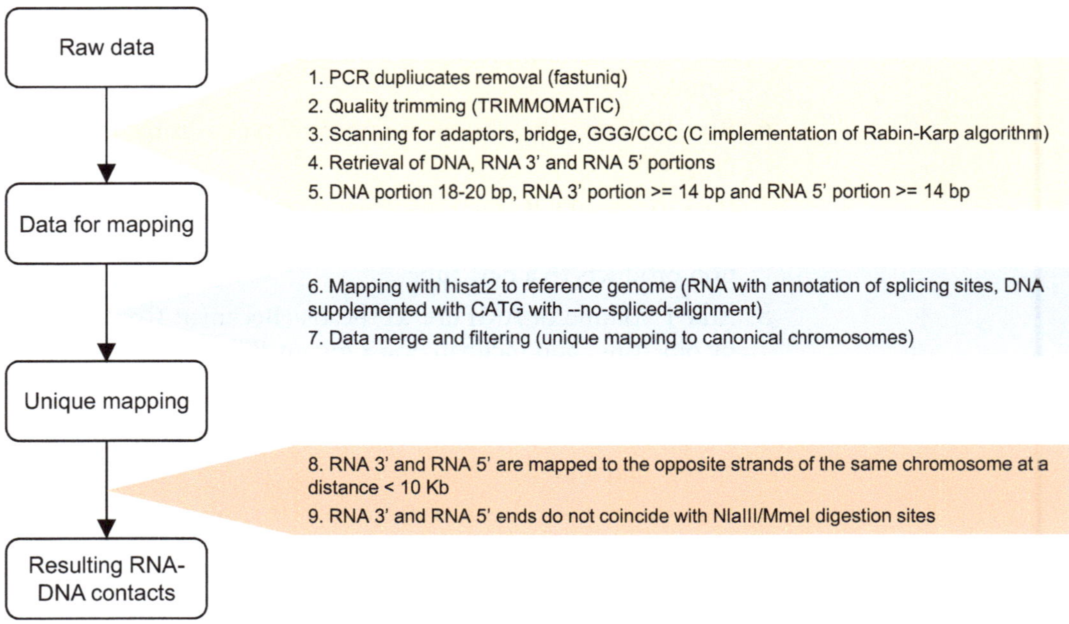

Fig. 3 Flowchart of the RedClib computational pipeline used for the processing of sequencing data generated in RedChIP experiments

Karp algorithm implementation in C. If a complete 37-mer bridge is identified in R1 read and R2 read starts with GGG, the read pair is retained.

4. Retrieval of DNA, RNA 3′, and RNA 5′ portions. The DNA portion is extracted as the region of R1 read upstream of the bridge (Fig. 1c). The RNA 3′ portion is extracted as the region of R1 read downstream of the bridge. If the RNA portion is short and R1 read extends to or beyond the switch template oligo, the RNA 3′ portion is extracted as the region of R1 read between the bridge and the switch template oligo (Fig. 1c). The RNA 5′ portion is extracted as the region of R2 read downstream of the first GGG. If the RNA portion is short and R2 read extends to or beyond the bridge, the RNA 5′ portion is extracted as the region of R2 read between the first GGG and the bridge (Fig. 1c). Note that RNA 3′ and RNA 5′ portions can partly or completely overlap if the RNA portion is short. We retain DNA portions of 18–20 nucleotides, RNA 3′ portions of ≥14 nucleotides, and RNA 5′ portions of ≥14 nucleotides (*see* **Notes 25** and **26**).
5. Mapping. DNA portions, RNA 3′ portions, and RNA 5′ portions are independently mapped to a reference genome with the hisat2 program. Before mapping, the end of the DNA portion adjoining the bridge is supplemented with CATG (the 3′ overhang produced by NlaIII digestion and then blunted, *see* Fig. 1a) to increase the yield of unique mappings. We use parameters: "-k 100 --no-spliced-alignment --no-softclip" for DNA and "-k 100 --no-softclip --dta-cufflinks" for RNAs (--known-splicesite-infile). SAM files are filtered for unique mappings with at most two mismatches relative to the reference genome with samtools and converted to BED with bedtools. We retain only such DNA–RNA 3′-RNA 5′ triples that were all successfully and uniquely mapped to the canonical chromosomes. If one of the portions is missing, non-uniquely mapped, unmapped, or mapped to a noncanonical chromosome, the read pair is filtered out.
6. Filtering out RNA portions whose ends colocalize with the NlaIII and MmeI digestion sites (*see* **Note 27**).
7. Filtering out reverse transcriptase template switching artifacts (*see* **Note 28**).

3.19 RNA Annotation

We use RNA 3′ portions retrieved from the forward reads as described in **step 4** of Subheading 3.18. If a splicing junction is reported within the RNA 3′ portion, we use a fragment of the RNA 3′ portion from the bridge to the break in alignment. We intersect RNA 3′ portions with a gene annotation for the genome assembly used and assign RNA 3′ portions to genes by the featureCounts program. We require that the RNA 3′ portion be mapped to the strand opposite to that of the gene as expected from the RedChIP procedure (Fig. 1a). Finally, we combine DNA portions mated with RNA 3′ portions originating from a single gene, thus obtaining a whole-genome contact profile for each respective RNA in the IP and input fractions.

3.20 Enrichment Calculation

For each RNA, we determine the number of contacts this RNA establishes with a genomic region of interest. Next, we summarize the contact numbers obtained for individual RNAs to calculate the total number of contacts established by all RNAs with the selected genomic region. Then, we divide the number of contacts of individual RNAs by the total number of contacts for all RNAs, thus yielding a representation of each RNA among all RNAs. The procedure is done separately for the IP and input fractions. Finally, we calculate the fold enrichment of RNA by dividing its representation in the IP fraction by its representation in the input fraction.

4 Notes

1. The standard procedure of the synthesis of a 5′ adenylated oligonucleotide includes blocking its 3′ end. A special request should be made upon purchasing the rApp oligo that it be provided with 3′ OH.
2. RedChIP uses a higher number of cells compared to the regular Red-C [20] or other RNA–DNA ligation protocols [19, 26, 27], which is dictated by the incorporation of the immunoprecipitation step. Usually, no more than 1% of chimeric RNA–DNA molecules remain in the immunoprecipitated fraction, depending on the antibody used. The low yield of RNA–DNA chimeric molecules in the immunoprecipitated fraction makes it necessary to use a higher number of PCR cycles to prepare a library for sequencing, which results in an increased number of PCR duplicates.
3. If desired, the cell pellet can be frozen in liquid nitrogen and stored at −80 °C.
4. SDS treatment removes the cytoplasm and extracts from fixed nuclei RNA and proteins that have not been cross-linked (directly or indirectly) to DNA.
5. Increasing the salt concentration helps to get a more clearly visible nuclear pellet at the next centrifugation step.
6. Dividing the nuclei suspension into two aliquots aims to ensure the proper mixing of the suspension during the restriction enzyme digestion step.

7. We recommend using a restriction enzyme producing 3′ overhangs such as NlaIII. It is also possible to use other restriction enzymes; however, this may require changing the digestion buffer and blunting protocol (*see* **Note 10**).
8. For isolation of DNA from control samples, we recommend using a standard protocol including Proteinase K treatment, phenol–chloroform extraction, ethanol precipitation, and RNase A treatment. Nuclear pellets can be stored at −20 °C until the DNA isolation.

   The sizes of DNA fragments after NlaIII digestion usually range from 300 bp to 5 kb, which may seem somewhat high for a 4-bp cutter such as NlaIII. However, it should be taken into account that the digestion reaction is carried out on cross-linked chromatin rather than pure DNA and therefore is not 100% efficient. In addition, short fragments produced by NlaIII cleavage may be washed out of the nuclei. The DNA integrity control should run on a gel as a single high molecular weight band.
9. RNAs with a phosphate group at the 3′ end can occur naturally in the cell or result from accidental fragmentation of RNA during incubations at 37 °C in the presence of Mg++. The RNA 3′ P ends should be dephosphorylated to allow ligation to the 5′ rApp ends of the bridge adapter. Removal of 3′ phosphate groups is carried out using T4 Polynucleotide Kinase in the absence of ATP.
10. The protocol described is for blunting 3′ protruding DNA ends. If a restriction enzyme-producing blunt ends is used, the DNA blunting step should be skipped. If a restriction enzyme producing 5′ overhangs is used, the DNA blunting should be done by filling-in ends in a 400-μL reaction containing 1× NEBuffer 2, 0.25 mM each dNTP, and 12-μL Klenow (5 U/μL, NEB) for 1 h at 25 °C with shaking at 900 rpm.
11. Several washes after the dA-tailing reaction aim to remove the traces of dATP that may have an inhibitory effect on T4 RNA Ligase 2 (truncated) at the next RNA ligation step.
12. While preparing the master mix, add vigorous 50% PEG-8000 as the last component using a pipette tip with the very tip cut off. Mix carefully by pipetting up and down, making sure to remove any remnants of the 50% PEG-8000 from within the tip.
13. T4 RNA Ligase 2, truncated (T4 Rnl2tr) specifically ligates the pre-adenylated 5′ end of DNA or RNA to the 3′ OH end of RNA. The enzyme does not use ATP for ligation, and is even inhibited in the presence of ATP, but requires the pre-adenylated substrate.

14. In the presence of SDS, the nuclei get swollen and precipitate as a large translucent loose pellet. Care must be taken when removing the supernatant. If the supernatant is not clear, the centrifugation speed can be increased to 7500× g.
15. The washing steps after the RNA ligation reaction are necessary to wash off the non-ligated bridge adapter from the fixed nuclei.
16. As genomic DNA is fragmented with a restriction enzyme leaving 5′ P and the bridge is provided with 5′ P (Fig. 1b), the DNA ends should be ready for ligation. However, just in case, we include a T4 Polynucleotide Kinase step to phosphorylate those DNA ends that may have lost a phosphate group in the previous steps of RedChIP protocol.
17. We carry out sonication using a probe-type sonicator VirTis VirSonic 100 at high power (setting 15) with, typically, 10 cycles of 20-s "ON" followed by 40-s "OFF," while keeping the sample immersed in an ice bath.
18. The pellet may not be visible after isopropanol centrifugation. We recommend pouring off the supernatant, draining the tube by placing it upside down on absorbent paper, and aspirating the remaining drops using a tip connected to a vacuum while avoiding the wall where the pellet should be.
19. MmeI cuts DNA predominantly at 20 and 21 bp upstream of the recognition site with approximately a 60%/40% ratio; some cuts also occur at 19 bp. MmeI requires two recognition sites for efficient DNA cleavage. If only one recognition site is present, the cleavage may be incomplete. Adding a short dsDNA oligo having a recognition site for MmeI stimulates the cleavage of DNA substrates containing a single MmeI site [28].
20. Reverse transcription is initiated from the 3′ hydroxyl of the biotinylated strand of the bridge (Fig. 1a,b). Reverse transcriptase can use both RNA and DNA as a template and can also transcribe across RNA-DNA junctions. Reverse transcriptase first transcribes bridge DNA, followed by DNA–RNA junction, and then RNA (Fig. 1a). Upon reaching the 5′ end of the RNA, reverse transcriptase adds a few non-template nucleotides (predominantly dC) to the 3′ end of cDNA. This dC stretch pairs with rGrGrG of the switch template oligo, and reverse transcriptase continues replication using the switch template oligo as a template [29] (Fig. 1a). Atypical nucleotides isocytidine and isoguanine prevent secondary switching at the 5′ end of the switch template oligo [30].
22. One end of NN-adapter carries two random 3′ overhanging bases (i.e., 16 possible dinucleotides) to ligate with NN 3′ ends

produced by MmeI digestion. NN-adaptor is provided without 5′ P to avoid adaptor-to-adaptor ligation. As a result, a nick is left in the non-biotinylated strand (Fig. 1a).

21. We use a library preparation strategy where a sequencing index is introduced with a PCR primer. PCR results in a single-indexed library ready for sequencing on an Illumina platform (Fig. 1c).

23. Two rounds of 1× AMPure XP size selection ensure the removal of the free PCR primer and primer dimers and deplete the amplified library of products <200 bp containing short unmappable RNA portions. Such products are generated in higher amounts in RedChIP compared to Red-C due to the incorporation of the sonication step causing additional RNA fragmentation. At the same time, DNA breaks caused by sonication do not lead to the appearance of PCR products with shortened DNA portions. Indeed, DNA breaks introduced at a distance of ≥20 bp from the bridge will not interfere with MmeI digestion, and DNA portions of the correct size will be produced; DNA breaks introduced at a distance of <20 bp from the bridge or within the bridge will result in the failure of MmeI digestion, NN-adapter ligation, and RNA–DNA chimera amplification.

24. The number of identified unique RNA–DNA contacts depends on various factors, including the sequencing depth, the presence of PCR duplicates, and the percentage of unmapped and multiply mapped reads. We note that the number of PCR duplicates is usually higher for IP compared to input libraries, even if those libraries are prepared from the same quantity of RNA–DNA chimeras, which likely reflects the lower complexity of IP libraries.

25. Most DNA portions are 18–20 nucleotides long, with the length distribution precisely following the MmeI digestion pattern (*see* **Note 18**). A 1-bp shift (18–20 instead of 19–21) is due to the presence of A/T base pair between the DNA portion and the MmeI recognition site (Fig. 1a). Reads without the DNA portion are also observed. Their quantity usually does not exceed 10%. Reads with DNA portion of >20 nucleotides are even rarer. Reads without DNA portion and with DNA portion of >20 nucleotides are filtered out.

26. In contrast to DNA portions, RNA portions demonstrate a wide range of length distribution. Short RNA portions may represent short RNA species or originate as a result of occasional fragmentation of RNA during multiple incubations at 37 °C in the presence of Mg++. We usually discard reads with RNA portion of <14 nucleotides to increase the percentage of uniquely mapped RNA portions.

27. We found that the end of the RNA 3′ portion adjoining the bridge and the end of the RNA 5′ portion adjoining GGG, which demarcate, respectively, the 3′ and 5′ ends of RNA in the RNA–DNA chimera, are slightly more frequently mapped to NlaIII sites than may be expected based on random distribution [20]. This observation indicates that some of the RNA–DNA chimeras (a small fraction) may actually be DNA–DNA chimeras produced by some aberrant ligation event and then used by reverse transcriptase for DNA-directed cDNA synthesis. We thus discard a read pair if 3′, 5′, or both ends of the RNA map within the NlaIII site ±1 letter. We also discard a read pair if the 5′ end of the RNA maps within the MmeI digestion site because the MmeI site may naturally occur within a NlaIII fragment. We realize that, concomitantly, we lose a fraction of genuine RNA–DNA ligation products because an RNA end may occasionally happen within the NlaIII and MmeI digestion sites. However, this is the trade-off for eliminating possible ligation artifacts.
28. To avoid artifacts that could originate from intermolecular template switching of the reverse transcriptase [31], we require that RNA 3′ and RNA 5′ portions be mapped to the opposite strands of the same chromosome at a distance of no more than 10 kb from each other. A threshold of 10 kb ensures the elimination of the vast majority (>99.9%) of possible template switching artifacts, while not affecting the detectability of genuine (continuous) RNA portions whose length is expected to be far below 10 kb based on the size distribution of PCR products (Fig. 2). If an RNA portion contains a splicing junction, the distance between the mapping coordinates of the RNA 5′ and 3′ ends may be higher than 10 kb. In order to retain such RNA portions, the distance threshold can be increased to 100 kb or higher, which still allows removing the majority of template-switching artifacts. Template switching artifacts are usually detected in a small fraction of reads (<1%).

Acknowledgments

This work was supported by the Russian Science Foundation (Grant 21-64-00001-P).

References

1. Engreitz JM, Ollikainen N, Guttman M (2016) Long non-coding RNAs: spatial amplifiers that control nuclear structure and gene expression. Nat Rev Mol Cell Biol 17(12): 756–770. https://doi.org/10.1038/nrm.2016.126
2. Sun Q, Hao Q, Prasanth KV (2018) Nuclear long noncoding RNAs: key regulators of gene

expression. Trends Genet 34(2):142–157. https://doi.org/10.1016/j.tig.2017.11.005

3. Li X, Fu XD (2019) Chromatin-associated RNAs as facilitators of functional genomic interactions. Nat Rev Genet 20(9):503–519. https://doi.org/10.1038/s41576-019-0135-1
4. Nozawa RS, Gilbert N (2019) RNA: nuclear glue for folding the genome. Trends Cell Biol 29(3):201–211. https://doi.org/10.1016/j.tcb.2018.12.003
5. Zong D, Oberdoerffer P, Batista PJ, Nussenzweig A (2020) RNA: a double-edged sword in genome maintenance. Nat Rev Genet 21(11): 651–670. https://doi.org/10.1038/s41576-020-0263-7
6. Razin SV, Gavrilov AA (2021) Non-coding RNAs in chromatin folding and nuclear organization. Cell Mol Life Sci 78(14):5489–5504. https://doi.org/10.1007/s00018-021-03876-w
7. Yeo SJ, Ying C, Fullwood MJ, Tergaonkar V (2023) Emerging regulatory mechanisms of noncoding RNAs in topologically associating domains. Trends Genet 39:217. https://doi.org/10.1016/j.tig.2022.12.003
8. Guo Q, Shi X, Wang X (2021) RNA and liquid-liquid phase separation. Noncoding RNA Res 6(2):92–99. https://doi.org/10.1016/j.ncrna.2021.04.003
9. Sharp PA, Chakraborty AK, Henninger JE, Young RA (2022) RNA in formation and regulation of transcriptional condensates. RNA 28(1):52–57. https://doi.org/10.1261/rna.078997.121
10. Chu C, Spitale RC, Chang HY (2015) Technologies to probe functions and mechanisms of long noncoding RNAs. Nat Struct Mol Biol 22(1):29–35. https://doi.org/10.1038/nsmb.2921
11. Chu C, Qu K, Zhong FL, Artandi SE, Chang HY (2011) Genomic maps of long noncoding RNA occupancy reveal principles of RNA-chromatin interactions. Mol Cell 44(4): 667–678. https://doi.org/10.1016/j.molcel.2011.08.027
12. Simon MD, Wang CI, Kharchenko PV, West JA, Chapman BA, Alekseyenko AA, Borowsky ML, Kuroda MI, Kingston RE (2011) The genomic binding sites of a noncoding RNA. Proc Natl Acad Sci USA 108(51): 20497–20502. https://doi.org/10.1073/pnas.1113536108
13. Engreitz JM, Pandya-Jones A, McDonel P, Shishkin A, Sirokman K, Surka C, Kadri S, Xing J, Goren A, Lander ES, Plath K, Guttman M (2013) The Xist lncRNA exploits three-dimensional genome architecture to spread across the X chromosome. Science 341(6147):1237973. https://doi.org/10.1126/science.1237973
14. Lieberman-Aiden E, van Berkum NL, Williams L, Imakaev M, Ragoczy T, Telling A, Amit I, Lajoie BR, Sabo PJ, Dorschner MO, Sandstrom R, Bernstein B, Bender MA, Groudine M, Gnirke A, Stamatoyannopoulos J, Mirny LA, Lander ES, Dekker J (2009) Comprehensive mapping of long-range interactions reveals folding principles of the human genome. Science 326(5950):289–293
15. Denker A, de Laat W (2016) The second decade of 3C technologies: detailed insights into nuclear organization. Genes Dev 30(12): 1357–1382. https://doi.org/10.1101/gad.281964.116
16. Sridhar B, Rivas-Astroza M, Nguyen TC, Chen W, Yan Z, Cao X, Hebert L, Zhong S (2017) Systematic mapping of RNA-chromatin interactions in vivo. Curr Biol 27(4):602–609. https://doi.org/10.1016/j.cub.2017.01.011
17. Li X, Zhou B, Chen L, Gou LT, Li H, Fu XD (2017) GRID-seq reveals the global RNA-chromatin interactome. Nat Biotechnol 35(10):940–950. https://doi.org/10.1038/nbt.3968
18. Bell JC, Jukam D, Teran NA, Risca VI, Smith OK, Johnson WL, Skotheim JM, Greenleaf WJ, Straight AF (2018) Chromatin-associated RNA sequencing (ChAR-seq) maps genome-wide RNA-to-DNA contacts. eLife 7:e27024. https://doi.org/10.7554/eLife.27024
19. Bonetti A, Agostini F, Suzuki AM, Hashimoto K, Pascarella G, Gimenez J, Roos L, Nash AJ, Ghilotti M, Cameron CJF, Valentine M, Medvedeva YA, Noguchi S, Agirre E, Kashi K, Samudyata, Luginbuhl J, Cazzoli R, Agrawal S, Luscombe NM, Blanchette M, Kasukawa T, Hoon M, Arner E, Lenhard B, Plessy C, Castelo-Branco-G, Orlando V, Carninci P (2020) RADICL-seq identifies general and cell type-specific principles of genome-wide RNA-chromatin interactions. Nat Commun 11(1):1018. https://doi.org/10.1038/s41467-020-14337-6
20. Gavrilov AA, Zharikova AA, Galitsyna AA, Luzhin AV, Rubanova NM, Golov AK, Petrova NV, Logacheva MD, Kantidze OL, Ulianov SV, Magnitov MD, Mironov AA, Razin SV (2020) Studying RNA-DNA interactome by Red-C identifies noncoding RNAs associated with various chromatin types and reveals

transcription dynamics. Nucleic Acids Res 48(12):6699–6714. https://doi.org/10.1093/nar/gkaa457

21. Gavrilov AA, Sultanov RI, Magnitov MD, Galitsyna AA, Dashinimaev EB, Lieberman Aiden E, Razin SV (2022) RedChIP identifies noncoding RNAs associated with genomic sites occupied by Polycomb and CTCF proteins. Proc Natl Acad Sci USA 119(1): e2116222119. https://doi.org/10.1073/pnas.2116222119
22. Mumbach MR, Rubin AJ, Flynn RA, Dai C, Khavari PA, Greenleaf WJ, Chang HY (2016) HiChIP: efficient and sensitive analysis of protein-directed genome architecture. Nat Methods 13(11):919–922. https://doi.org/10.1038/nmeth.3999
23. Fang R, Yu M, Li G, Chee S, Liu T, Schmitt AD, Ren B (2016) Mapping of long-range chromatin interactions by proximity ligation-assisted ChIP-seq. Cell Res 26(12): 1345–1348. https://doi.org/10.1038/cr.2016.137
24. Gavrilov A, Razin SV, Cavalli G (2015) In vivo formaldehyde cross-linking: it is time for black box analysis. Brief Funct Genomics 14(2): 163–165. https://doi.org/10.1093/bfgp/elu037
25. Hendrickson GD, Kelley DR, Tenen D, Bernstein B, Rinn JL (2016) Widespread RNA binding by chromatin-associated proteins. Genome Biol 17:28. https://doi.org/10.1186/s13059-016-0878-3
26. Zhou B, Li X, Luo D, Lim DH, Zhou Y, Fu XD (2019) GRID-seq for comprehensive analysis of global RNA-chromatin interactions. Nat Protoc 14(7):2036–2068. https://doi.org/10.1038/s41596-019-0172-4
27. Wu W, Yan Z, Nguyen TC, Bouman Chen Z, Chien S, Zhong S (2019) Mapping RNA-chromatin interactions by sequencing with iMARGI. Nat Protoc 14(11): 3243–3272. https://doi.org/10.1038/s41596-019-0229-4
28. Morgan RD, Bhatia TK, Lovasco L, Davis TB (2008) MmeI: a minimal Type II restriction-modification system that only modifies one DNA strand for host protection. Nucleic Acids Res 36(20):6558–6570. https://doi.org/10.1093/nar/gkn711
29. Zhu YY, Machleder EM, Chenchik A, Li R, Siebert PD (2001) Reverse transcriptase template switching: a SMART approach for full-length cDNA library construction. BioTechniques 30(4):892–897. https://doi.org/10.2144/01304pf02
30. Kapteyn J, He R, McDowell ET, Gang DR (2010) Incorporation of non-natural nucleotides into template-switching oligonucleotides reduces background and improves cDNA synthesis from very small RNA samples. BMC Genomics 11:413. https://doi.org/10.1186/1471-2164-11-413
31. Houseley J, Tollervey D (2010) Apparent non-canonical trans-splicing is generated by reverse transcriptase in vitro. PLoS One 5(8): e12271. https://doi.org/10.1371/journal.pone.0012271

Chapter 2

Mapping of RBP-Mediated RNA–RNA Interactome with RIC-Seq

Zhaokui Cai and Yuanchao Xue

Abstract

RNA–RNA interactions tend to be mediated by RNA-binding proteins (RBPs) and are involved in every step of gene regulation. Faithfully mapping RBP-mediated intra- and intermolecular RNA–RNA interactions is a prerequisite for simultaneously inferring noncoding RNA structures and targets. We present a recently developed RIC-seq (RNA in situ conformation sequencing) technology for global profiling of RBP-mediated in situ RNA–RNA interactions. Our approach fixes RBP-mediated RNA–RNA interactions in living cells with formaldehyde for subsequent permeabilization, micrococcal nuclease fragmentation, pCp-biotin labeling, and in situ RNA–RNA proximity ligation. These proximally ligated chimeric RNA fragments are subsequently enriched and converted into pair-end libraries for deep sequencing. Aligning these chimeric fragments to the genome allows the identification of the whole complement of RBP-mediated RNA–RNA interactome in cells and tissue samples. Combining these proximity information and bioinformatic tools enables the unbiased study of RNA structures and targets.

Key words RNA, RNA interactions, RIC-seq, RNA-binding protein, Proximity ligation

1 Introduction

Gene expression can be controlled at multiple levels, such as transcriptional activation or repression, co-transcriptional and posttranscriptional processing. Noncoding RNAs regulate gene expression during these processes by functioning as guides, scaffolds, and decoys [1]. To achieve these regulations, noncoding RNAs, especially long noncoding RNAs, usually need to fold intricate tertiary structures via intramolecular base pairing and interact with other RNA molecules through intermolecular base pairing [2, 3]. All these intra- and intermolecular RNA–RNA interactions need RNA-binding protein (RBP) to stabilize or sometimes as a mediator [4, 5]. Therefore, systematically mapping RBP-mediated RNA–RNA interactions becomes a prerequisite for deciphering the structure and target of noncoding RNAs.

Michiaki Hamada et al. (eds.), *RNA-RNA and RNA-DNA Interactions*, Methods in Molecular Biology, vol. 2949,
https://doi.org/10.1007/978-1-0716-4670-0_2,

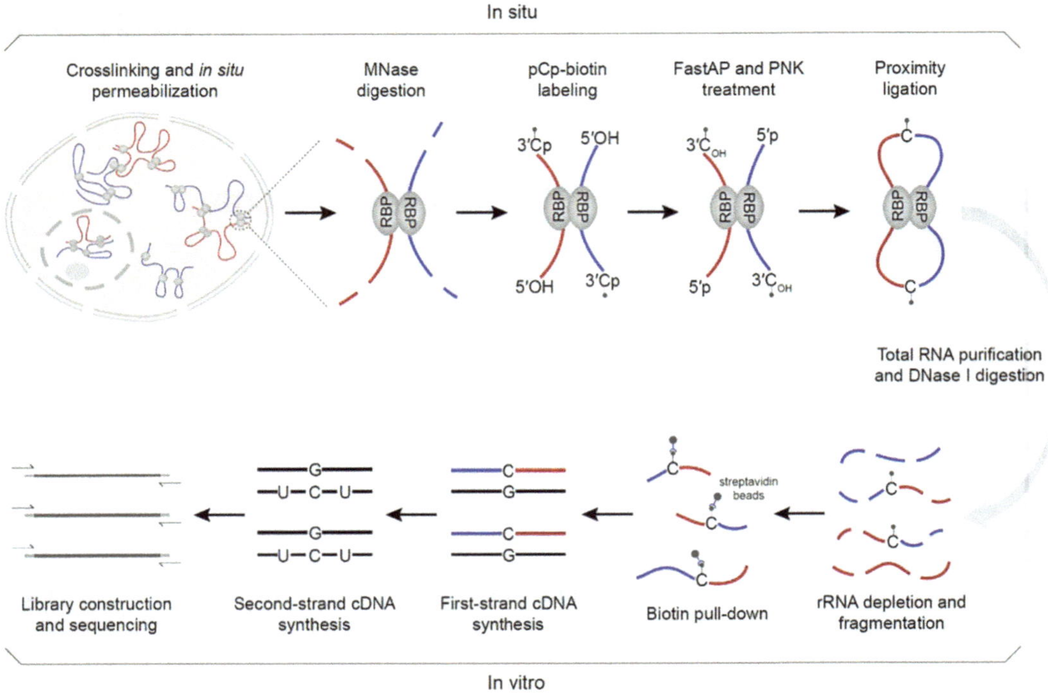

Fig. 1 Schematic diagram of RIC-seq technology. The in situ part of RIC-seq includes cross-linking, permeabilization, MNase digestion, pCp-biotin labeling, and proximity ligation. The in vitro part includes rRNA depletion, RNA fragmentation, biotin-labeled RNA enrichment, first-strand and second-strand cDNA synthesis, strand-specific library construction, and sequencing. Abbreviations: MNase micrococcal nuclease, pCp-biotin biotinylated 3′,5′-bisphosphate cytidine, and RBP RNA-binding protein

Many state-of-the-art methods have been developed to map protein-mediated RNA–RNA interactions, such as CLASH, hiCLIP, RIPPLiT, RPL, and MARIO [6–10]. These methods have significantly advanced our understanding of RBP-mediated RNA–RNA interactions in gene regulation. However, CLASH, hiCLIP, and RIPPLiT methods must ectopically express a specific RBP [6–8], which may perturb native RNA–RNA interaction networks in living cells. In addition, these three methods perform RNA–RNA proximity ligation in vitro in a diluted solution and thus may introduce many spurious ligations. Although RPL and MARIO methods intend to profile all the RBP-associated RNA–RNA interactions [9, 10], the lack of cross-linking in the RPL method or over-cross-linking in the MARIO approach limited their applications.

To address these limitations, we recently developed an RNA in situ conformation sequencing technology RIC-seq (Fig. 1) for global profiling of the whole complement of RBP-mediated RNA–RNA interactome [3, 11]. Several technical advancements are introduced: (1) RIC-seq uses formaldehyde to fix RBP-mediated RNA–RNA in situ interactions in a short time to reduce nonspecific cross-linking adducts; (2) pCp-biotin is applied to label the 3′ end of proximally interacting RNA fragments for marking the chimeric junctures, which enables efficient discrimination of pairwise interacting RNA fragments; and (3) proximity ligation is performed in situ rather than in vitro, thus significantly reducing the potential spurious ligations in a diluted solution (Fig. 1). These three technical advancements significantly increased our approach's efficiency, resolution, and signal-to-noise ratio in detecting protein-mediated RNA–RNA interactome.

2 Materials

2.1 Reagents

1. RNaseZap™ RNase decontamination solution.
2. Diethyl pyrocarbonate (DEPC).
3. PBS.
4. Formaldehyde (37%, w/v).
5. Glycine.
6. Tris.
7. Hydrochloric acid (HCl).
8. Magnesium chloride ($MgCl_2$).
9. Ethylene glycol-bis (2-aminoethylether)-N,N,N′,N′--tetraacetic acid (EGTA).
10. Ethylenediamine tetraacetic acid (EDTA).
11. IGEPAL® CA-630 (also known as NP-40).
12. Triton X-100.
13. Tween 20.
14. Protease inhibitors.
15. SUPERase In.
16. Micrococcal nuclease (MNase).
17. Calcium chloride ($CaCl_2$).
18. FastAP thermosensitive alkaline phosphatase (Fast AP).
19. Pierce™ RNA 3′ End Biotinylation Kit.
20. Adenosine triphosphate (ATP, 10 mM).
21. T4 polynucleotide kinase (T4 PNK, Thermo Fisher Scientific).
22. T4 RNA ligase.
23. RiboLock RNase inhibitor.
24. Proteinase K.
25. Sodium dodecyl sulfate (SDS).
26. Trizol LS.
27. Chloroform.

28. Isopropanol.
29. Glycoblue.
30. RQ1 RNase-Free DNase.
31. Sodium acetate solution (3 M, pH 5.2).
32. Ethanol.
33. RNase H.
34. TURBO DNase.
35. RNA clean and concentrator-5 kit.
36. Qubit™ RNA High Sensitivity (HS) Assay Kit.
37. Dynabeads™ MyOne™ Streptavidin C1.
38. Sodium hydroxide (NaOH).
39. Sodium chloride (NaCl).
40. Yeast RNA.
41. Acid–phenol: Chloroform (pH 4.5).
42. SuperScript™ II reverse transcriptase.
43. dNTP mix (10 mM).
44. Second-strand buffer.
45. dATP (100 mM).
46. dGTP (100 mM).
47. dCTP (100 mM).
48. dTTP (100 mM).
49. dUTP (100 mM).
50. *Escherichia coli* DNA polymerase I.
51. AMPure XP beads.
52. dNTP mix (25 mM).
53. T4 DNA polymerase.
54. Klenow fragment.
55. T4 polynucleotide kinase (enzymatics).
56. Klenow (3′–5′ exo-).
57. T4 DNA ligase (rapid).
58. Platinum SuperFi DNA polymerase.
59. USER enzyme.
60. Biowest agarose.
61. GelRed.
62. MinElute Gel Extraction Kit.
63. Qubit™ DNA High Sensitivity (HS) Assay Kit.
64. Nuclease-free water (not DEPC-treated).

2.2 Buffers (See Note 1)

1. 1× PNK buffer (0.2% v/v NP-40): 50 mM Tris–HCl pH 7.4, 10 mM $MgCl_2$, 0.2% (v/v) NP-40.
2. 1× PNK buffer (0.05% v/v NP-40): 50 mM Tris–HCl pH 7.4, 10 mM $MgCl_2$, 0.05% (v/v) NP-40.
3. 1× PNK + EGTA buffer: 50 mM Tris–HCl pH 7.4, 20 mM EGTA, 0.5% (v/v) NP-40
4. High-salt wash buffer: 5× PBS (no Mg^{2+}, Ca^{2+}), 0.5% (v/v) NP-40.
5. Permeabilization buffer: 10 mM Tris–HCl pH 7.5, 10 mM NaCl, 0.5% (v/v) NP-40, 0.3% (v/v) Triton X-100, 0.1% (v/v) Tween 20, 1× protease inhibitors, 2 U/mL SUPERase In.
6. 1× MN reaction buffer: 50 mM Tris–HCl pH 8.0, 5 mM $CaCl_2$.
7. Proteinase K buffer: 10 mM Tris–HCl pH 7.5, 10 mM EDTA, 0.5% (w/v) SDS.
8. 5× Hybridization buffer: 1 M NaCl, 500 mM Tris–HCl pH 7.4.
9. Solution A: 100 mM NaOH, 50 mM NaCl.
10. Solution B: 100 mM NaCl.
11. 2× TWB buffer: 10 mM Tris–HCl pH 7.5, 1 mM EDTA, 2 M NaCl, 0.02% Tween 20.
12. PK buffer: 100 mM NaCl, 10 mM Tris–HCl pH 7.0, 1 mM EDTA, 0.5% (w/v) SDS.
13. 1× TBE: 890 mM Tris, 890 mM boric acid, 20 mM EDTA.

2.3 Oligos and Primers

1. PEI Adapter oligo A: 5′ /Phos/ GATCGGAAGAGCA CACGTCT 3′.
2. PEI Adapter oligo B: 5′ ACACTCTTTCCCTACACGACGC TCTTCCGATCT 3′.
3. P5 primer: AATGATACGGCGACCACCGAGATCTACACT CTTTCCCTACACG.
   ACGCTCTTCCGATCT.
4. P7 Index primer: CAAGCAGAAGACGGCATACGAGATA NNNNNNNNGTGAC.
   TGGAGTTCAGACGTGTGCTCTTCCGATCT
   ("NNNNNN" denotes the 8 bp randomized barcode).

2.4 Equipment

1. Freezer (2–8 °C, −25 °C, and − 80 °C).
2. 5 and 10 mL disposable serological pipette.
3. 10 and 15 cm cell culture dish.
4. Cell culture incubator.

5. 15 and 50 mL centrifuge tube.
6. 0.22 μm filter.
7. PH meter.
8. Cell scrapers.
9. 1.5 mL LoBind microcentrifuge tubes.
10. Eppendorf ThermoMixer® C.
11. 0.2 mL PCR tubes.
12. DynaMag™-2 Magnet.
13. Sub-cell GT Horizontal Electrophoresis Systems.
14. Gel tray.
15. Safe ImagerTM 2.0 blue-light transilluminator.
16. Razor blade.
17. Vortex mixers.
18. Tube rotator.
19. NanoDrop 2000c.
20. Qubit® 3.0 fluorometer.
21. Refrigerated centrifuge and microcentrifuge.
22. Single-channel manual pipettes (0.1–2 μL, 2–20 μL, 20–200 μL, 100–1000 μL).
23. Thermal cycler.
24. Agilent 2100 bioanalyzer.

3 Methods

3.1 Adapter Preparation

1. Add 10 μL of PEI adapter oligo A (100 μM) and 10 μL of PEI adapter oligo B (100 μM) to a 0.2 mL PCR tube and mix well by pipetting up and down.
2. Incubate the tube at 70 °C for 10 min in a thermal cycler. Place the tube on the laboratory bench and let the mixture cool to room temperature naturally. The concentration of the annealed adapter is 50 μM.
3. Dilute the annealed adapter to 2 μM with nuclease-free water for the subsequent RIC-seq experiment. The diluted adapter can be aliquoted and stored at −20 °C for 1 year.

3.2 Cell Crosslinking

1. Discard the cell culture medium in 10-cm dishes and wash cells three times with 10-mL ice-cold PBS (*see* **Note 2**).
2. Add 10 mL of freshly prepared 1% formaldehyde in PBS solution to the washed cells and incubate at room temperature for 10 min to fix the RBP-mediated in situ RNA–RNA interaction.

3. Add 500 μL of 2.5 M glycine directly to the dish to quench the formaldehyde by gently mixing in a rocking platform at room temperature for 10 min.
4. Discard the quenched solution and wash cells three times with 10-mL ice-cold PBS.
5. Scrape the cells from the dish with a cell scraper and transfer the cell suspension to a 50 mL centrifuge tube. Pellet down the cells by centrifuging at 1200× *g* for 10 min at 4 °C (*see* **Notes 3** and **4**).
6. Discard the supernatant and resuspend the pellet in 1 mL of ice-cold PBS, then transfer the cell suspension to a new 1.5 mL Eppendorf tube.
7. Centrifuge at 600× *g* for 10 min at 4 °C and discard the supernatant. Cell pellets can be processed immediately in the next step or stored at −80 °C for no more than one week.

3.3 Cell Permeabilization and RNA Fragmentation

1. Resuspend the cell pellet in 1-mL permeabilization buffer. Mix well by pipetting up and down ten times.
2. Incubate the cell suspension on ice for 15 min and mix by inverting every 2 min.
3. Collect the cell pellet by centrifuging at 1200× *g* for 5 min at 4 °C and discard the supernatant.
4. Wash the cell pellet three times with ice-cold 1× PNK buffer (0.2% v/v NP-40). For each washing step, resuspend the pellet with 600 μL of ice-cold 1× PNK buffer (0.2% v/v NP-40), mix well by pipetting up and down five times, rotate at 20 rpm on a rotator for 5 min at 4 °C, centrifuge at 1200× *g* for 5 min at 4 ° C, and discard the supernatant. Please note that in the following steps, the cell pellets are washed in the same manner as described in this step.
5. Dilute MNase 1:1000 with 1× MN buffer. Mix well and dilute the above diluent 1:10 with 1× MN buffer (*see* **Note 5**).
6. Resuspend the cell pellet with 200 μL of diluted MNase and mix gently with a cut tip by pipetting up and down ten times.
7. Incubate the mixture for 10 min at 37 °C in a ThermoMixer C with 15 s of intermittent mixing at 1000 rpm every 2 min.
8. Collect the cell pellet by centrifuging at 1200× *g* for 5 min at 4 °C and discard the supernatant.
9. Wash cells twice with 600 μL of ice-cold 1× PNK + EGTA buffer, then twice with 600 μL of ice-cold 1× PNK buffer. Discard the supernatant after the last wash.

3.4 pCp-Biotin Labeling (See Note 6)

1. Set up the Fast AP reaction mixture as follows:

10× Fast AP buffer	10 μL
FastAP thermosensitive alkaline phosphatase	10 μL
Nuclease-free water	80 μL
Total	100 μL

2. Resuspend the cell pellet with the Fast AP reaction mixture and mix gently with a cut tip by pipetting up and down ten times.
3. Incubate the mixture for 15 min at 37 °C in a ThermoMixer C with 15 s of intermittent mixing at 1000 rpm every 2 min.
4. Collect the cell pellet by centrifuging at 1200× *g* for 5 min at 4 °C and discard the supernatant.
5. Wash cells twice with 600 μL of ice-cold 1× PNK + EGTA buffer, twice with 600 μL of ice-cold high-salt buffer, and three times with 600 μL of ice-cold 1× PNK buffer (0.05% v/v NP-40). Discard the supernatant thoroughly after the last wash.
6. Set up the labeling reaction mixture without 30% (w/v) PEG20000 as follows (*see* **Note 7**):

10× reaction buffer	10 μL
Ribolock RNase inhibitor	6 μL
Biotinylated Cytidine (Bis) phosphate	4 μL
T4 RNA ligase	10 μL
Nuclease-free water	20 μL
Total	50 μL

7. Resuspend the cell pellet with the labeling reaction mixture and mix gently with a cut tip by pipetting up and down ten times.
8. Add 50 μL of 30% (w/v) PEG 20,000 using a cutoff 200-μL tip to the cell suspension and mix gently by pipetting up and down ten times.
9. Incubate the mixture overnight at 16 °C in a ThermoMixer C with 15 s of intermittent mixing at 1200 rpm every 3 min (*see* **Note 8**).
10. Add 2 μL of T4 RNA ligase and 4 μL of 10 mM ATP to the reaction mixture the next day. Mix gently with a cut tip by pipetting up and down ten times.
11. Incubate the mixture for 3 h at 16 °C in a ThermoMixer C with 15 s of intermittent mixing at 1200 rpm every 3 min.

12. Collect the cell pellet by centrifuging at 1200× *g* for 5 min at 4 °C and discard the supernatant.
13. Wash cells three times with 600 μL of ice-cold 1× PNK buffer (0.2% v/v NP-40). Discard the supernatant thoroughly after the last wash.

3.5 Proximity Ligation

1. Set up the Fast AP reaction mixture as follows:

10× Fast AP buffer	10 μL
FastAP thermosensitive alkaline phosphatase	10 μL
Nuclease-free water	80 μL
Total	100 μL

2. Resuspend the cell pellet with the Fast AP reaction mixture and mix gently with a cut tip by pipetting up and down ten times.
3. Incubate the mixture for 15 min at 37 °C in a ThermoMixer C with 15 s of intermittent mixing at 1000 rpm every 3 min.
4. Collect the cell pellet by centrifuging at 1200× *g* for 5 min at 4 °C and discard the supernatant.
5. Wash cells twice with 600 μL of ice-cold 1× PNK + EGTA buffer, twice with 600 μL of ice-cold high-salt buffer, and three times with 600 μL of ice-cold 1× PNK buffer (0.2% v/v NP-40). Discard the supernatant thoroughly after the last wash.
6. Set up the T4 PNK reaction mixture as follows:

10× reaction buffer A	10 μL
T4 polynucleotide kinase	10 μL
10 mM ATP	15 μL
Nuclease-free water	65 μL
Total	100 μL

7. Resuspend the cell pellet with the T4 PNK reaction mixture and mix gently with a cut tip by pipetting up and down ten times.
8. Incubate the mixture for 45 min at 37 °C in a ThermoMixer C with 15 s of intermittent mixing at 1000 rpm every 3 min.
9. Collect the cell pellet by centrifuging at 1200× *g* for 5 min at 4 °C and discard the supernatant.
10. Wash cells twice with 600 μL of ice-cold 1× PNK + EGTA buffer and twice with 600 μL of ice-cold 1× PNK buffer (0.05%

v/v NP-40). Discard the supernatant thoroughly after the last wash.

11. Set up the proximity ligation reaction mixture as follows:

10× reaction buffer	20 μL
Ribolock RNase inhibitor	8 μL
T4 RNA ligase	10 μL
BSA (1 mg/mL)	20 μL
Nuclease-free water	142 μL
Total	200 μL

12. Resuspend the cell pellet with the proximity ligation reaction mixture and mix gently with a cut tip by pipetting up and down ten times.
13. Incubate the mixture overnight at 16 °C in a ThermoMixer C with 15 s of intermittent mixing at 1000 rpm every 3 min.
14. Add 2 μL of T4 RNA ligase and 4 μL of 10 mM ATP to the reaction mixture the next day. Mix gently with a cut tip by pipetting up and down ten times.
15. Incubate the mixture for 3 h at 16 °C in a ThermoMixer C with 15 s of intermittent mixing at 1000 rpm every 3 min.
16. Collect the cell pellet by centrifuging at 1200× *g* for 5 min at 4 °C and discard the supernatant.
17. Wash cells three times with 600 μL of ice-cold 1× PNK buffer (0.2% v/v NP-40). Discard the supernatant thoroughly after the last wash.

3.6 RNA Purification

1. Add 50 μL of proteinase K and 200 μL of proteinase K buffer to the cell pellet. Mix gently by pipetting up and down ten times.
2. Incubate the mixture for 60 min at 37 °C in a ThermoMixer C with 15 s of intermittent mixing at 1000 rpm every 3 min, then 56 °C for 15 min. After the reaction, let the mixture cool to room temperature.
3. Add 750 μL of Trizol LS to the mixture and homogenize the sample by pipetting up and down ten times. Incubate the sample at room temperature for 5 min.
4. Add 220 μL of chloroform to the mixture and shake vigorously for 15 s. Incubate the sample at room temperature for 3 min.
5. Centrifuge the tube at 16,000× *g* for 10 min at 4 °C and transfer the aqueous phase to a new 1.5-mL Eppendorf tube.

6. Add 1 μL of GlycoBlue and 500 μL of isopropanol to the tube and mix gently by pipetting up and down ten times. Place the tube at −20 °C overnight.
7. Next day, centrifuge the tube at 16,000× *g* for 15 min at 4 °C. Discard the supernatant and wash the RNA pellet twice with 600 μL of 75% (v/v) ethanol.
8. Centrifuge the tube at 16,000× *g* for 5 min at 4 °C. Discard the supernatant thoroughly and air-dry the pellet for 2 min.
9. Dissolve the pellet in 20 μL of nuclease-free water. Measure the concentration of the RNA sample using NanoDrop 2000c.
10. The expected total RNA yield ranges from 50 to 80 μg. Take 20 μg total RNA for subsequent experiments. The remaining RNA samples can be stored at −80 °C.

3.7 DNase I Treatment

1. Set up the DNase I reaction mixture as follows:

Total RNA (20 μg)	x μL
10× RQ1 DNase reaction buffer	10 μL
Ribolock RNase inhibitor	3 μL
RQ1 RNase-free DNase	5 μL
Nuclease-free water	82-x μL
Total	100 μL

2. Incubate the mixture for 20 min at 37 °C in a ThermoMixer C.
3. Add 100 μL of nuclease-free water and 200 μL of acid phenol: chloroform to the sample. Mix well by shaking.
4. Centrifuge the tube at 16,000× *g* for 10 min at 4 °C and transfer the aqueous phase to a new 1.5-mL Eppendorf tube.
5. Add 20 μL of 3 M NaOAc, 1 μL of glycoblue, and 500 μL of 100% (v/v) ethanol to the sample. Mix gently by pipetting up and down ten times. Place the tube at −20 °C overnight.
6. Next day, centrifuge the tube at 16,000× *g* for 15 min at 4 °C. Discard the supernatant and wash the RNA pellet twice with 600 μL of 75% (v/v) ethanol.
7. Centrifuge the tube at 16,000× *g* for 5 min at 4 °C. Discard the supernatant thoroughly and air-dry the pellet for 2 min.
8. Dissolve the pellet in 6 μL of nuclease-free water.
9. Take 0.2 μL of the RNA sample and dilute 1:20 with nuclease-free water. Detect the length distribution of RNA fragments by the Agilent 2100 Bioanalyzer.

3.8 rRNA Depletion (See Note 9)

1. Set up the hybridization reaction mixture as follows:

RNA	5.8 μL
rRNA probe mix (2 μg/μL)	10 μL
5× hybridization buffer	4 μL
Nuclease-free water	0.2 μL
Total	20 μL

2. Mix gently by pipetting up and down 20 times. Incubate the sample at 95 °C for 2 min in a thermal cycler, then gradually slow down to 22 °C at a rate of −0.1 °C/s, hold at 22 °C for 5 min, and immediately place the tubes on ice.
3. Set up the RNase H reaction mixture as follows:

RNA/rRNA probe mix	20 μL
RNase H	5 μL
10× RNase H buffer	3 μL
Nuclease-free water	2 μL
Total	30 μL

4. Mix gently by pipetting up and down 20 times. Incubate the sample at 37 °C for 30 min and immediately place the tubes on ice.
5. Set up the DNase reaction mixture as follows:

Last step sample	30 μL
Turbo DNase	5 μL
10× Turbo DNase buffer	4 μL
Nuclease-free water	1 μL
Total	40 μL

6. Mix gently by pipetting up and down 20 times. Incubate the sample at 37 °C for 30 min and immediately place the tubes on ice.
7. Add 10 μL of nuclease-free water to the tube and transfer the sample to a new 1.5-mL Eppendorf tube. Purify RNA with the RNA Clean and Concentrator-5 kit following the manufacturer's instructions. Finally, elute the RNA with 18.5 μL of nuclease-free water. Typically, users can obtain a 17 μL sample. Save 1 μL of RNA and quantify it using a Qubit fluorometer

and Qubit RNA HS Assay Kit following the manufacturer's instructions. The expected concentration of the RNA product should be 100–200 ng/μL (*see* **Note 10**). Transfer the remaining sample to a 0.2-mL PCR tube.

3.9 pCp-Biotin Selection

1. Add 4 μL of 5× first-strand buffer to the tube and mix gently by pipetting up and down 20 times.
2. Incubate the sample at 94 °C for 5 min and immediately place the tubes on ice.
3. 20 μL of Dynabeads MyOne Streptavidin C1 beads should be used for each sample. Based on your number of samples, transfer the appropriate amount of C1 beads to a new 1.5-mL Eppendorf tube (*see* **Note 11**).
4. Place the tube on a magnetic rack for 1 min and discard the supernatant.
5. Add an equal volume of solution A to the tube to resuspend the C1 beads and mix well by pipetting up and down ten times.
6. Incubate the sample for 2 min at room temperature.
7. Place the tube on a magnetic rack for 1 min and discard the supernatant.
8. Wash C1 beads once more by repeating **steps 5–7**.
9. Add an equal volume of solution B to the tube to resuspend the C1 beads and mix well by pipetting up and down ten times.
10. Place the tube on a magnetic rack for 1 min and discard the supernatant.
11. Add 32 μL of yeast RNA (50 μg), 68 μL of DEPC water, and 100 μL of 2× TWB buffer to the tube. Mix gently by pipetting up and down 20 times.
12. Place the tube on a rotator and incubate the sample at 20 rpm for 60 min at room temperature.
13. After instantaneous centrifugation on the IKA MiniG Centrifuges for 2 s, place the tube on a magnetic rack for 1 min and discard the supernatant. Wash the beads three times with 600 μL of 1× TWB buffer.
14. Place the tube on a magnetic rack for 1 min and discard the supernatant. Add the RNA sample from **step 2**, 30 μL of nuclease-free water, and 50 μL of 2× TWB buffer to the tube. Mix gently by pipetting up and down ten times.
15. Place the tube on a rotator and incubate the sample at 20 rpm for 30 min at room temperature.
16. After instantaneous centrifugation, place the tube on a magnetic rack for 1 min and discard the supernatant. Wash the beads four times with 600 μL of 1× TWB buffer.

17. Place the tube on a magnetic rack for 1 min and discard the supernatant.
18. Resuspend the beads with 100 μL of PK buffer and mix gently by pipetting up and down ten times.
19. Incubate the sample for 10 min at 95 °C in a ThermoMixer C with continuous mixing at 1000 rpm.
20. Place the tube on a magnetic rack for 1 min after instantaneous centrifugation. Transfer the supernatant to a new 1.5-mL Eppendorf tube.
21. Repeat **steps 18–20** once more to elute RNA thoroughly.
22. Rinse beads briefly with 100 μL of PK buffer and place the tube on a magnetic rack for 1 min. Transfer the supernatant to the Eppendorf tube, and the total volume of elute should be 300 μL.
23. Add 300 μL of Acid Phenol: Chloroform to the tube and mix well by shaking.
24. Centrifuge the tube at 16,000× *g* for 10 min at 4 °C and transfer the aqueous phase to a new 1.5-mL Eppendorf tube.
25. Add 18 μL of 5 M NaCl, 1 μL of glycoblue, and 900 μL of 100% (v/v) ethanol to the sample. Mix gently by pipetting up and down ten times. Place the tube at −20 °C overnight.
26. Next day, centrifuge the tube at 16,000× *g* for 15 min at 4 °C. Discard the supernatant and wash the RNA pellet twice with 600 μL of 75% (v/v) ethanol.
27. Centrifuge the tube at 16,000× *g* for 5 min at 4 °C. Discard the supernatant thoroughly and air-dry the pellet for 2 min.
28. Dissolve the pellet in 10 μL of nuclease-free water and transfer the RNA sample to a new 0.2-mL PCR tube.

3.10 Strand-Specific Library Construction

3.10.1 First-Strand cDNA Synthesis

1. Add 0.5 μL of 0.1 μg/μL N6 primer to the sample and mix well by pipetting up and down 20 times.
2. Incubate the sample at 65 °C for 5 min in a thermal cycler, then immediately place the tubes on ice for at least 2 min.
3. Set up the first-strand cDNA synthesis reaction mixture as follows:

5× first-strand buffer	3 μL
10 mM dNTPs	1 μL
Ribolock RNase inhibitor	0.5 μL
0.1 M DTT	0.5 μL
Superscript II reverse transcriptase	0.5 μL
Total	5.5 μL

4. Add the above mixture to the RNA sample from **step 2** and mix well by pipetting up and down 20 times.
5. Place the tube in a thermal cycler. Incubate the sample for 10 min at 25 °C, 40 min at 42 °C, and 15 min at 70 °C; then hold on at 4 °C.

3.10.2 Second-Strand DNA Synthesis

1. Set up the second-strand DNA synthesis reaction mixture as follows:

5× second-strand buffer	10 μL
25 mM dNTPs (dUTP, *see* **Note 12**)	0.8 μL
RNase H	0.2 μL
Ultrapure water	20.5 μL
E. Coli DNA polymerase I	2.5 μL
Total	34 μL

2. Transfer the first-strand cDNA product to a new 1.5-mL Eppendorf tube. Add the above mixture to the tube and mix well by pipetting up and down 20 times.
3. Incubate the mixture for 2 h at 16 °C in a ThermoMixer C with 15 s of intermittent mixing at 300 rpm every 3 min.
4. Equilibrate the aliquoted AMPure beads at room temperature for 30 min in advance—vortex for 30 s to homogenize the beads.
5. Add 90 μL of AMPure beads to the sample from **step 3** and mix well by pipetting up and down ten times.
6. Incubate the sample for 5 min at room temperature. Place the tube on a magnetic rack until the solution becomes clear.
7. Discard the supernatant. Add 200 μL of freshly prepared 80% (v/v) ethanol to the tube while keeping the tube on the magnetic rack and gently rotate the tube for 2 weeks.
8. Wash beads once more by repeating **step 7**.
9. Discard the ethanol thoroughly and let the tube open on the magnetic rack to air-dry for 1–3 min.
10. Add 44 μL of Qiagen Elution Buffer to the tube to resuspend the beads, mix well by pipetting up and down ten times, and incubate the sample for 5 min at room temperature.
11. Place the tube on a magnetic rack until the solution becomes clear. Transfer the supernatant (43 μL) to a new 1.5-mL Eppendorf tube.
12. Take 1 μL of the sample and measure the concentration using a Qubit fluorometer and Qubit dsDNA HS Assay Kit following the manufacturer's instructions. The expected concentration of the DNA product should be 0.2–0.6 ng/μL.

3.10.3 End Repair

1. Set up the end repair reaction mixture as follows:

10× T4 polynucleotide kinase buffer	5 μL
25 mM dNTPs	0.4 μL
T4 DNA polymerase	1.2 μL
Klenow fragment	0.2 μL
T4 polynucleotide kinase	1.2 μL
Total	8 μL

2. Add the above mixture to the sample to purify the second-strand DNA product, mix well by pipetting up and down ten times, and incubate the sample in a ThermoMixer C for 30 min at 20 °C.
3. Purify the end-repaired DNA product with 90 μL of AMPure beads as described in Subheading 3.10.2.
4. Elute DNA from AMPure beads with 20.5 μL of Qiagen Elution Buffer and transfer the product (19.7 μL) to a new 1.5-mL Eppendorf tube.

3.10.4 dA-Tailing

1. Set up the dA-tailing reaction mixture as follows:

10× blue buffer	2.3 μL
5 mM dATP	0.5 μL
Klenow (3′–5′ exo-)	0.5 μL
Total	3.3 μL

2. Add the above mixture to the end-repaired DNA product, mix well by pipetting up and down ten times, and incubate the sample in a ThermoMixer C for 30 min at 37 °C.

3.10.5 Adapter Ligation

1. Set up the adapter ligation reaction mixture without T4 DNA ligase (Rapid) as follows:

10× rapid ligation buffer	1.4 μL
10 mM ATP	0.1 μL
Adapter (2 μM, *see* **Note 13**)	1 μL
Total	2.5 μL

2. Add the above mixture to the last step sample, and mix well by pipetting up and down ten times.

3. Add 1 μL of T4 DNA ligase (Rapid) to the sample, mix well by pipetting up and down ten times, and incubate the sample in a ThermoMixer C for 15 min at 20 °C (*see* **Note 14**).
4. Purify the adapter-ligated DNA product with 47.7 μL of AMPure beads as described in Subheading 3.10.2.
5. Elute DNA from AMPure beads with 25.5 μL of Qiagen Elution Buffer and transfer the product (25 μL) to a new 1.5-mL Eppendorf tube.
6. Purify the adapter-ligated DNA product with 45 μL of AMPure beads as described in Subheading 3.10.2.
7. Elute DNA from AMPure beads with 17 μL of Qiagen Elution Buffer and transfer the product (16.5 μL) to a new 0.2-mL PCR tube.

3.10.6 PCR Amplification

1. Choosing the appropriate PCR cycle number to reduce PCR duplication is critical for a successful RIC-seq experiment. We usually take 2 μL of the purified sample from the last step to test the optimal PCR cycle number by amplifying with two different cycles, such as 10 and 14 (*see* **Note 15**). Set up the PCR reaction mixture in a 0.2-mL PCR tube as follows:

Adapter-ligated DNA	1 μL
5× SuperFi buffer	5 μL
25 mM dNTPs	0.4 μL
P5 primer (10 μM)	1 μL
P7 index primer (10 μM)	1 μL
Platinum SuperFi DNA polymerase	0.1 μL
User enzyme	0.5 μL
Ultrapure water	16 μL
Total	25 μL

2. Mix well by pipetting up and down 20 times and perform the following program for PCR amplification:

37 °C	15 min
98 °C	30 s
10 and 14 cycles:	
98 °C	10 s
62 °C	30 s
72 °C	30 s

(continued)

Final extension:	
72 °C	5 min
12 °C	Hold on

3. Add 5 μL of 6× Orange G loading buffer to the tube and mix well by pipetting up and down ten times. Perform 2% agarose gel electrophoresis in 1× TBE at 120 V for ~1 h.
4. Check the smear band with a blue-light transilluminator.
5. Use the optical PCR cycle number for final amplification according to the signal intensity of the smear band. Set up the PCR reaction mixture in a 0.2-mL PCR tube as follows:

Adapter-ligated DNA	14.2 μL
5× SuperFi buffer	5 μL
25 mM dNTPs	0.4 μL
P5 primer (10 μM)	1 μL
P7 index primer (10 μM, *see* **Note 16**)	1 μL
Platinum SuperFi DNA polymerase	0.4 μL
User enzyme	3 μL
Total	25 μL

6. Mix well by pipetting up and down 20 times and perform the following program for PCR amplification:

37 °C	15 min
98 °C	30 s
Appropriate cycles:	
98 °C	10 s
62 °C	30 s
72 °C	30 s
Final extension:	
72 °C	5 min
12 °C	Hold on

7. Add 5 μL of 6× Orange G loading buffer to the tube and mix well by pipetting up and down ten times. Perform 2% agarose gel electrophoresis in 1× TBE at 120 V for ~1 h.
8. Check the smear band with a blue-light transilluminator (Fig. 2, *see* **Note 17**).

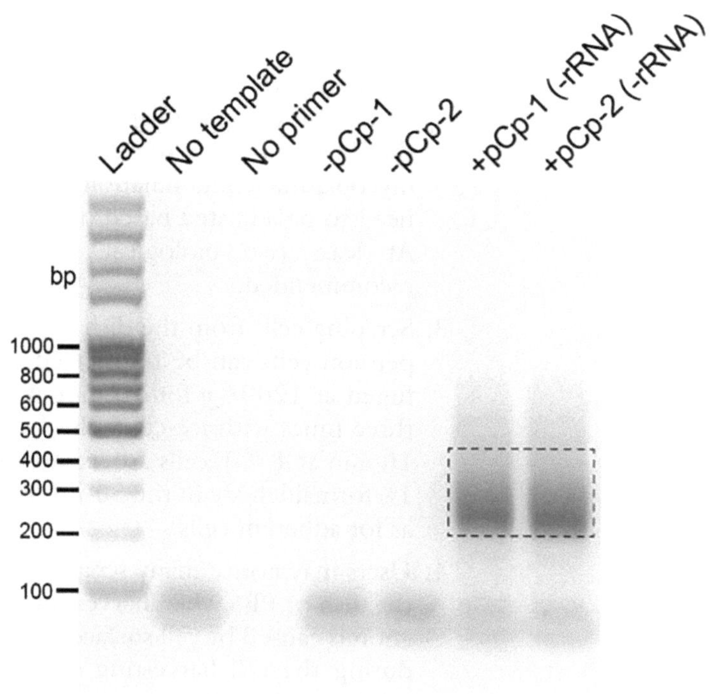

Fig. 2 Fragment distribution of RIC-seq libraries showed on an agarose gel. -pCp is a negative control, denoting that the sample is not labeled with pCp-biotin. The smear band from 200 to 450 bp in the dashed box is purified from gel and sequenced on the NovaSeq 6000 System

9. Cut the smeared band between 200 and 450 bp and extract the DNA product with a Qiagen MinElute Gel Extraction Kit following the manufacturer's instructions.
10. Elute the DNA with 16.5 μL of Elution Buffer. Typically, users can obtain 15 μL of product. Take 1 μL of the sample and measure the concentration using a Qubit fluorometer and Qubit dsDNA HS Assay Kit following the manufacturer's instructions. The expected DNA yield should be 20–100 ng.

4 Notes

1. All the tubes and tips used in the RIC-seq experiment should be RNase- and DNase-free. All the buffers used in the RIC-seq experiment should be filtered with a 0.22 μm PVDF membrane filter. To avoid RNase contamination, the bench, pipettes, and other equipment should be treated with RNAseZAP® before the RIC-seq experiment. Users should wear a lab coat, mask,

and gloves throughout the experiment and change gloves frequently.

2. A typical RIC-seq experiment requires approximately 1×10^7 HeLa cells to yield an approximate 50–70 μL cell pellet. Users should ensure that the cells used for RIC-seq are free from mycoplasma contamination. The starting cell amount may need to be adjusted based on the cell size and RNA contents. At least two biological replicates for each sample are recommended.

3. Scraping cells from the dish is suitable for adherent cells. Suspension cells can be transferred to a 50-mL tube and centrifuged at 1200× *g* for 10 min at 4 °C directly. After washing three times with ice-cold PBS by centrifugation (1200× *g* for 10 min at 4 °C), cells are fixed with 10 mL of PBS containing 1% formaldehyde in the 50-mL tube using the same condition as for adherent cells.

4. Users may notice many scraped cells floating on the surface of the dish or PBS when harvesting adherent cells. This phenomenon is caused by cell surface tension, which can cause cell loss during the cell harvesting step. Users can add 0.01% (v/v) NP-40 to the cell suspension to reduce surface tension.

5. Dilution of the MNase at 1:10,000 is suitable for most human and mouse cell lines. Users may need to titrate an appropriate concentration of MNase when performing RIC-seq experiments with tissues or other organisms (e.g., plant samples).

6. When performing pCp-biotin labeling, users can set up a negative control without pCp-biotin by replacing it with nuclease-free water to evaluate the specificity of the chimeric enrichment.

7. PEG 20,000 (30%) is very viscous. Aliquot this reagent into the cell suspension using cutoff 200 μL tips. Homogenize the cell suspension containing 30% PEG 20,000 by pipetting gently and avoiding vortexing.

8. Cell suspension containing 30% PEG 20,000 is viscous. The pCp-labeling should be performed at 16 °C overnight in a ThermoMixer C with 15 s of intermittent mixing at 1200 rpm every 3 min to avoid cells sinking to the bottom of the tube.

9. Ribosome RNA (rRNA) should be removed from total RNA using an RNase H-based rRNA depletion method. Commercial kits are available for such purposes, such as the Ribo-off rRNA depletion kit (Vazyme, N406) or NEBNext® rRNA Depletion Kit (Human/Mouse/Rat). However, the kit cost is too high due to the large amount of RNA input in the RIC-seq experiment. To this end, we followed a published RNase H-

based protocol [12] to synthesize 195 50-nt DNA probes that cover the entire rRNA and reverse complement to each rRNA (28S rRNA, 18 s rRNA, 5.8S rRNA, 16S rRNA, 12SrRNA, and 5S rRNA). All probes are mixed at equal molar amounts and used to deplete rRNA.

10. Ribosome RNA usually accounts for about 90% of total RNA. Users can estimate the depletion efficiency by the amount of RNA obtained after rRNA removal. Users can repeat the rRNA depletion steps if the RNA yield exceeds the expected amount.
11. Since it takes about 1.5 h to block the C1 beads, users can save experimental time by preparing the C1 beads after the Turbo DNase treatment step has started.
12. dNTP mixture used in second-strand DNA synthesis contains 25 mM dATP, 25 mM dGTP, 25 mM dCTP, 5 mM dTTP, and 20 mM dUTP. To prepare the dNTP solution, we typically mix 50 μL of 100 mM dATP, 50 μL of 100 mM dGTP, 50 μL of 100 mM dCTP, 10 μL of 100 mM dTTP, and 40 μL of 100 mM dUTP thoroughly. The dNTP solution can be aliquoted and stored at −20 °C.
13. The yield of DNA after second-strand DNA synthesis determines the amount of adapter used. If the yield of DNA is below or above the range in Subheading 3.10.2, users can adjust the amount of adapters used according to the amount of DNA and the average length of the fragment.
14. Users should ensure that T4 DNA (rapid) ligase is added to the mixture last and strictly control the reaction time to reduce the adapter's self-ligation.
15. We typically test PCR cycle numbers with 10 and 14 cycles. Users can adjust the PCR cycle number based on DNA yield after second-strand DNA synthesis.
16. When manipulating multiple samples, users should use different index primers in this step which is critical for differentiating pooled sequencing samples.
17. To avoid DNA damage, users should use blue-light transilluminators instead of UV transilluminators.

References

1. Chen J, Xue Y (2016) Emerging roles of non-coding RNAs in epigenetic regulation. Sci China Life Sci 59:227–235
2. Wang D, Ye R, Cai Z, Xue Y (2022) Emerging roles of RNA-RNA interactions in transcriptional regulation. Wiley Interdiscip Rev RNA 13:e1712
3. Cai Z, Cao C, Ji L, Ye R, Wang D, Xia C, Wang S, Du Z, Hu N, Yu X, Chen J, Wang L, Yang X, He S, Xue Y (2020) RIC-seq for global in situ profiling of RNA-RNA spatial interactions. Nature 582:432–437
4. Van Treeck B, Parker R (2018) Emerging roles for intermolecular RNA-RNA interactions in RNP assemblies. Cell 174:791–802
5. Hentze MW, Castello A, Schwarzl T, Preiss T (2018) A brave new world of RNA-binding proteins. Nat Rev Mol Cell Biol 19:327–341

6. Kudla G, Granneman S, Hahn D, Beggs JD, Tollervey D (2011) Cross-linking, ligation, and sequencing of hybrids reveals RNA-RNA interactions in yeast. Proc Natl Acad Sci USA 108:10010–10015
7. Sugimoto Y, Vigilante A, Darbo E, Zirra A, Militti C, D'Ambrogio A, Luscombe NM, Ule J (2015) hiCLIP reveals the in vivo atlas of mRNA secondary structures recognized by Staufen 1. Nature 519:491–494
8. Metkar M, Ozadam H, Lajoie BR, Imakaev M, Mirny LA, Dekker J, Moore MJ (2018) Higher-order organization principles of pre-translational mRNPs. Mol Cell 72:715–726. e713
9. Ramani V, Qiu R, Shendure J (2015) High-throughput determination of RNA structure by proximity ligation. Nat Biotechnol 33: 980–984
10. Nguyen TC, Cao X, Yu P, Xiao S, Lu J, Biase FH, Sridhar B, Huang N, Zhang K, Zhong S (2016) Mapping RNA-RNA interactome and RNA structure in vivo by MARIO. Nat Commun 7:12023
11. Cao C, Cai Z, Ye R, Su R, Hu N, Zhao H, Xue Y (2021) Global in situ profiling of RNA-RNA spatial interactions with RIC-seq. Nat Protoc 16:2916–2946
12. Adiconis X, Borges-Rivera D, Satija R, DeLuca DS, Busby MA, Berlin AM, Sivachenko A, Thompson DA, Wysoker A, Fennell T, Gnirke A, Pochet N, Regev A, Levin JZ (2013) Comparative analysis of RNA sequencing methods for degraded or low-input samples. Nat Methods 10:623–629

Chapter 3

Integrated Analysis of Cross-Link-Ligation Data for the Detection of RNA 2D/3D Structures and Interactions In Vivo

Wilson H. Lee, Minjie Zhang, and Zhipeng Lu

Abstract

The latest advancements in cross-linking- and proximity ligation-based methods have transformed the study of RNA structures and interactions in living cells. Despite this progress, sequencing data must be meticulously processed to extract the rich information. This presents distinct challenges, as noted in recent methods, e.g., Psoralen Analysis of RNA Interactions and Structures (PARIS) and Spatial 2′-Hydroxyl Acylation Reversible Cross-linking (SHARC). In this chapter, we detail the protocol used to analyze RNA sequencing data derived from cross-link-ligation experiments. We outline the strategies used to preprocess raw data, map to a reference genome, classify reads, and assemble duplex groups. Further, we show an example of analyzing structural data for the ribosomal RNA.

Key words Bioinformatics, Cross-linking, Next generation sequencing, Proximity ligation, RNA structure, RNA, RNA interaction

1 Introduction

As one of the principal informational biomolecules, RNA is a fundamental component to various macromolecular machineries [1, 2]. Over the last few decades, the classical belief of RNA functioning as merely a mediator between DNA genetic information and functional protein products has been replaced by the understanding that RNA is a versatile biomolecule with diverse functions [3–6]. RNAs not only engage in various aspects of gene regulation, protein synthesis, and posttranscriptional modifications, but have also been found to play dynamic roles in cell signaling and other intra- and intercellular interactions [7–10]. For instance, the multitude of noncoding RNAs (ncRNAs) in cells, including small and large RNAs, have been shown to interact with other ncRNAs, messenger RNAs (mRNAs), proteins, and small molecules [11–15].

Michiaki Hamada et al. (eds.), *RNA-RNA and RNA-DNA Interactions*, Methods in Molecular Biology, vol. 2949,
https://doi.org/10.1007/978-1-0716-4670-0_3,

Although the complexity of RNA is increasingly appreciated, the structures and interactions of RNA in cells remain poorly understood [8, 16]. The secondary and tertiary structures of many known RNAs are still in an early stage of characterization, although tools such as high-throughput sequencing and computational analysis have expedited the discovery of new RNAs [17, 18]. Manipulation and observation of RNA properties is critical to structure elucidation, although it has long been arduous and technically challenging. Physical methods such as X-ray crystallography, nuclear magnetic resonance (NMR) spectroscopy, and cryo-electron microscopy (EM) have made substantial progress in solving 3D structures of RNAs and their associated complexes [19–23]. In parallel, computational approaches, such as calculations of minimum free energy conformations or conservation of structural motifs in homologous nucleotide sequences, have been used to assist RNA secondary or tertiary structure prediction [24–26]. While these approaches yield structures of few small RNAs, they remain time-consuming and unsuitable for in vivo analyses due to various sample type limitations (e.g., size, flexibility, shape, etc.) [27, 28].

Efforts in the development of next generation sequencing-based RNA structure probing methods have enabled studies of RNA in cells [29]. These methods combine various enzymes and chemical agents, followed by sequencing in high throughput. Broadly, they can be grouped into either footprinting- or proximity ligation-based approaches.

Footprinting-based methods use probes to modify RNA nucleotides or cleave the chain, which are then captured by reverse transcription and detected by sequencing and downstream analysis [30, 31]. Examples such as dimethyl sulfate (DMS) footprinting and 2′-hydroxyl acylation analyzed by primer extension (SHAPE) have been previously used to probe the flexibility and reactivity of individual nucleotides [32–34]. These chemical methods result in an indirect one-dimensional information, useful for guiding structure prediction and providing structural context to the nucleotides.

More recently, cross-linking-based methods, e.g., CLASH, COMRADES, LIGR-seq, PARIS, SPLASH, and SHARC, have been used to interrogate RNA structural features more directly, revealing spatial proximity between fragments [8, 35–42]. These methods typically employ chemical reagents that react with pairs of RNA nucleotides in spatial proximity. Upon fragmentation and purification/enrichment of cross-linked RNA fragments, the fragment ends are enzymatically ligated and sequenced in high throughput to yield a chimeric read, wherein segments come from distinct regions or entirely different RNA molecules.

Although the advent of the cross-linking-based methods has deepened our understanding of RNA structures and interactions, there remain major challenges in analyzing the data from these experimental techniques. Factors such as random fragmentation, complex cross-linking and proximity ligation orientation, and non-continuous reads convolute the data processing [43, 44]. To address these problems, it is important to standardize a pipeline for the systematic analysis of newly generated and previously published experimental data.

To date, myriad studies have reported the analysis of chimeric reads in sequencing data from different methods [29, 45–49]. Most of the applications of these computational tools have been limited to their respective studies [35, 36, 38]. As such, an integrated analytical approach that utilizes robust tools allows for a reproducible analysis across different sequencing data obtained in cross-link-ligation studies [43, 50]. In separate manuscripts, we presented detailed protocols for the experimental methods on PARIS and SHARC [51, 52]. In this chapter, we detail a stepwise pipeline for the analysis of cross-link-ligation derived experimental data using available software and scripts. We expect that sequencing data analyzed in a similar fashion will allow for insightful results, leading to a deeper understanding of RNA structures and interactions in living cells.

2 Materials

This chapter outlines a systematic approach for handling data derived from RNA cross-link-ligation studies (*see* Fig. 1a),

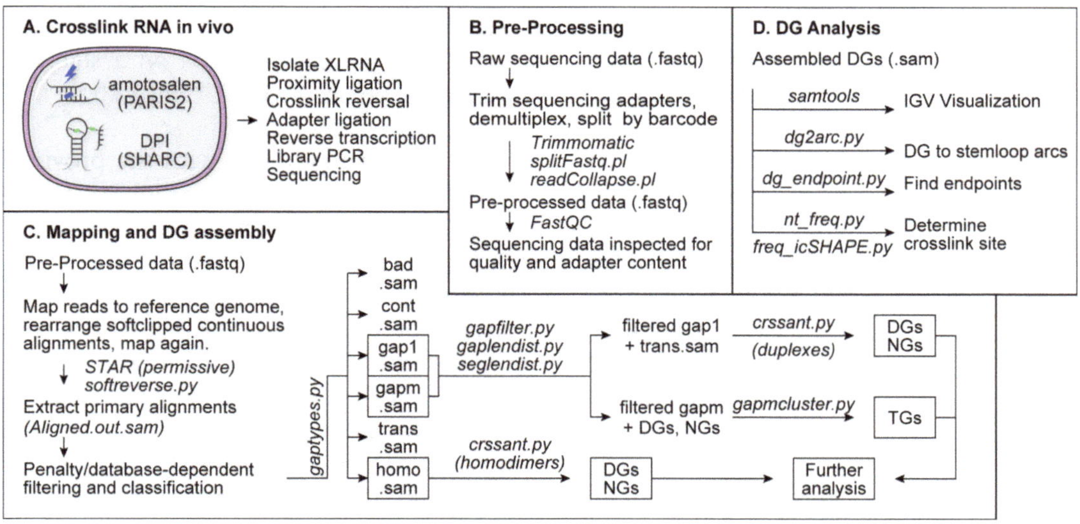

Fig. 1 Stepwise diagram of the post-sequencing data handling process for cross-link-ligation-based experiments. (**a**) Overview of the cross-linking and library preparation steps leading up to sequencing. (**b**) Preprocessing steps to clean up raw data, perform data quality control. (**c**) Flowchart of the mapping and processes leading to CRSSANT analysis. (**d**) Examples of further DG analysis

including sections for preprocessing, mapping to a reference, secondary structure analysis via CRSSANT (cross-linked RNA secondary structure analysis using network techniques), and further analysis. Steps from each section can be used independently and tailored to specific analytical purposes. Preprocessing, although described, is not strictly necessary for subsequent steps since data handling varies between library preparation approaches (i.e., ligation-based, tagmentation-based, amplicon-based) and choice of sequencing platform (*see* Fig. 1b). All reads must be mapped to a reference genome prior to CRSSANT steps and subsequent analysis (*see* Fig. 1c). Familiarity with Linux/Unix environments, command line interface, NGS short read technologies, and basic python and shell scripting skills is recommended.

2.1 Hardware

A high-performance compute (HPC) cluster with a 64-bit Unix-based operating system, x86–64 compatible processer, and at least 32 GB of RAM is recommended for completing this protocol. Optionally, preprocessing steps can be performed locally on a Unix-based system prior to mapping and CRSSANT, both of which require an HPC cluster with >30 GB memory. Post-mapping and CRSSANT analysis steps can be performed locally on a system with at least four compute threads and 8 GB of RAM. Note that large datasets require more memory and upward of 1 week for mapping and CRSSANT duplex group (DG) and nonoverlapping group (NG) assembly. Disk space requirements are dependent on the size of raw sequencing reads (fastq) and subsequent aligned and processed read (sam/bam) files. For reference, operations described for preprocessing and mapping steps were performed on an HPC cluster (2.60 GHz 8-core Intel Xeon 2640v3, 32 GB, CentOS) and DG analysis steps on a standard desktop (3.3 GHz 4-core Intel Core i5, 8 GB, Ubuntu 20.04).

2.2 Software

As described, the protocol can be run entirely with the provided scripts and open-source tools. For detailed information regarding installation and troubleshooting, consult the manual, reference, or website linked with each tool (*see* **Note 1**). Use the software versions as detailed in this pipeline for error-free operation. While both legacy and updated versions of programs are available, various parameters and features may vary. Additionally, it is highly recommended that users have a GUI-based text editor to manipulate files and script parameters. Test datasets and example output files are supplied for crucial steps of the protocol (*see* **Note 2**). To allow for maximum parameter control and troubleshooting, each step of the pipeline should be run separately. Sample shell scripts and representative data for a typical pipeline are also provided as an example (*see* **Note 3**).

2.2.1 Environment Setup

1. Python 3.x with packages biopython, matplotlib, NetworkX (v2.1+), numpy, pandas, pysam, scikit-learn, and SciPy. Note that Unix-based systems have a version of Python installed by default, but only versions 3.x + will work with scripts in this protocol. See https://www.python.org/ for detailed installation instructions. We recommend using the Anaconda/Bioconda package managers to download and install dependencies.

2.2.2 Preprocessing

1. Trimmomatic-0.36 [53]. http://www.usadellab.org/cms/uploads/supplementary/Trimmomatic/Trimmomatic-0.36.zip
2. icSHAPE pipeline [54]. https://github.com/qczhang/icSHAPE
3. FastQC [55]. https://www.bioinformatics.babraham.ac.uk/projects/fastqc/

2.2.3 Mapping to Reference Genome

1. STAR-2.7.1a [56]. https://github.com/alexdobin/STAR
   https://github.com/alexdobin/STAR/archive/refs/tags/2.7.1a.tar.gz
2. samtools v1.1+ [57, 58]. https://www.htslib.org/download/
```
$ cd samtools-1.x
$ make
$ make install
$ export PATH/bin:$PATH
```
3. Read classification and counting. https://github.com/whl-usc/rna2d3d/tree/main/tools

2.2.4 CRSSANT and DG/NG Analysis

1. bedtools v2.22+ [59, 60]. https://github.com/arq5x/bedtools2
2. CRSSANT [43]. https://github.com/zhipenglu/CRSSANT/archive/refs/heads/master.zip
3. DG_analysis. https://github.com/whl-usc/rna2d3d/tree/main/scripts
4. Integrative Genomics Viewer (IGV) [61, 62]. https://software.broadinstitute.org/software/igv/download

2.2.5 Directory Setup and File Naming Conventions

1. Prepare directories for software, raw data, and working paths (*see* Fig. 2). A directory containing relevant software (*see* Fig. 2a) and references (*see* Fig. 2b) will simplify access for future use (*see* **Note 4**). Preprocessed and downstream analysis files should be kept in directories separate from raw data (*see* Fig. 2c).
2. Keep file naming conventions consistent between analyses of different datasets. This simplifies any changes to shell and

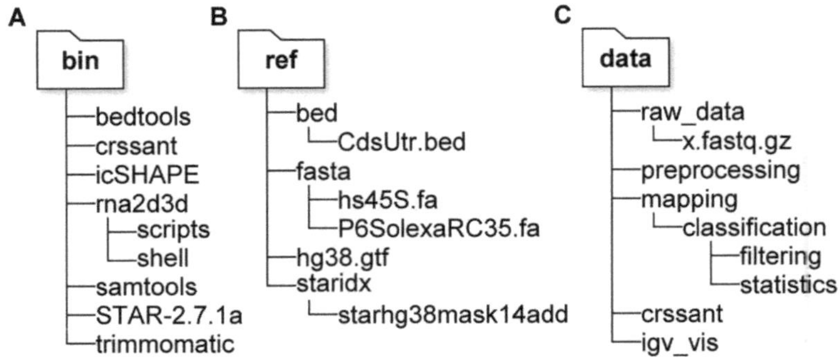

Fig. 2 Sample directory trees for files relevant to the analytical pipeline. Note that only names of key files and subdirectories are included for clarity. Refer to the corresponding sections and Table 1 for further details. Consolidated directories contain the following: (**a**) all relevant software and scripts, (**b**) reference sequences, genomes, and bed files, (**c**) data files, isolated into appropriate subdirectories

python scripts when processing multiple datasets and prevents files from being erroneously overwritten. Table 1 shows an example of filenames with sample name/prefix "x."

3 Methods

3.1 Data Preprocessing

In this section, post-sequencing data is preprocessed to remove sequencing adapters, short reads, PCR duplicates and demultiplexed by barcode, if applicable (*see* **Note 5**). Upon completion of these steps, sample sequencing data is processed and suitable for mapping to a reference genome (*see* **Note 6**).

3.1.1 Trimming, PCR Duplicate Removal, Splitting by Barcode

1. Run Trimmomatic in single-ended mode to trim the 3′-end adapter sequences. Use functions "ILLUMINACLIP" and "SLIDINGWINDOW" with 16 threads and base quality encoding phred33. Specify the input, output, and adapter sequence files (*see* **Notes 6** and 7).

```
$ java -jar trimmomatic-0.36.jar SE -threads 16 -phred33 x.fastq.gz x.trim3.fastq ILLUMINACLIP:P6SolexaRC35.fa:3:20:10 SLIDINGWINDOW:4:20 MINLEN:18
```

2. Remove PCR duplicates by using built-in random hexamers using script readCollapse (*see* **Note 8**).

```
$ perl readCollapse x_trim3.fastq x_trim3_nodup.fastq
```

3. Where applicable, split the multiplexed library by barcodes using splitFastq.pl (*see* **Note 9**). Use parameter "-U" to indicate single-ended reads for the input file, "-l" to specify a list of strings denoting different libraries, and "-b" to specify the barcode position and length. Optionally include "-s" to generate simple library statistics.

Table 1
Example file names and associated descriptions. Note that the naming convention as described in the table assumes multiple barcodes ("bcnn.fastq") are used and that demultiplexing is performed on the sample(s)

File names	Description
Preprocessing (trimming, demultiplexing, splitting by barcode)	
x.fastq.gz	Raw data (compressed)
x_trim3.fastq	3′-end adapter sequence removed
bcnn.fastq	Split library based on barcode and name (e.g., CGTGAT: RTR701 has "bcnn" RTR701)
x_trim3_nodup.fastq	PCR duplicates removed
x_trim_nodup_bcnn.fastq	5′-end adapter sequence removed
STAR mapping to genome (permissive)	
x_Aligned.sortedByCoord.out.bam	All aligned reads mapped to STAR index, sorted by coordinate
x_Aligned.out.bam	All aligned reads mapped to STAR index, unsorted and in BAM format
xLog.final.out	Overall summary of mapping statistics
xLog.out	Main log file, detailed run information
xLog.progress.out	Job progress statistics (e.g., processed read number, mapped read statistics, etc.)
SJ.out.tab	High confidence splice junctions
Unmapped.out.mate1	Output of unmapped or partially mapped reads
Extracting primary alignments	
x_nonchimeric_temp.sam	Reads not aligned to multiple strands or chromosomes
x_chimeric_temp.sam	Reads aligned to multiple strands or chromosomes
x_nonchimeric_pri.bam	Non-chimeric primary reads converted to BAM format, discards supplementary and secondary reads
x_nonchimeric_pri.sam	Non-chimeric primary reads converted to SAM format
x_pri.sam	Combined primary non-chimeric and chimeric reads
Classifying alignments	
x_cont.sam	Continuous alignments, no gaps
x_gap1.sam	Noncontinuous alignments, 1 gap
x_gapm.sam	Noncontinuous alignments, >1 gap
x_trans.sam	Noncontinuous alignments, 2 arms on different strands or chromosomes
x_homo.sam	Noncontinuous alignments, 2 arms overlap each other

(continued)

Table 1
(continued)

File names	Description
x_bad.sam	Noncontinuous alignments, complex combinations of indels and gaps
x_log.out	Log file for the run, details input and output alignment count
Filtering spliced and short gaps	
x_prigap1_filtered.sam	x_prigap1 filtered for splice junctions, short deletions
x_prigapm_filtered.sam	x_prigapm filtered for splice junctions, short deletions
x_prigaps_filtered.sam	Combined prigap1 and prigapm files for segment and gap statistics
Reads segment and gap statistics	
x_gaplen.pdf	PDF figure showing cumulative distribution histogram of the gap lengths
x_seglen.pdf	PDF figure showing cumulative distribution histogram of the segment lengths
x_pri_biotype.txt	Distribution of RNA into various types
x_pri_ReadCount.txt	Count of reads mapped to specific genes
x_pri_RPKM.txt	RPKM values for each gene
X_RPM.txt	RPM values for each gene
CRSSANT	
x_pri_crssant.sam	Combined SAM file containing both gap1 and trans reads from gaptypes.py
x_pri_crssant_TAG.sam	CRSSANT primary reads file tagged with unique identifier for separation after DG assembly
header	Sequencing file header from any SAM file associated with the same sample
x_pri_crssant.cliques.t_o0.2_dg.bedpe	Bedpe output file listing all duplex groups formed
x_pri_crssant.cliques.t_o0.2.sam	SAM file containing alignments assigned to DGS, with DG and NG annotations
Assembling DGs/NGs for rRNA	
x_prigap1_hs45S.tmp	Ribosomal RNA subunit hs45S extracted from all primary reads with 1 gap, with potentially SA tag present (column 21)
x_hs45S.sam	Ribosomal RNA subunit hs45S, SA removed, sequencing header added
x_hs45S_####.sam	Predefined number of randomly selected (e.g., "####") hs45S reads.
Data visualization with IGV	

(continued)

Table 1 (continued)

File names	Description
x.bam	Compressed binary representation of SAM file containing all primary reads
x_sorted.bam	BAM file sorted by reference coordinates
x_sorted.bam.bai	BAM file index, stores BAM file position

```
$ perl splitFastq.pl -U x_trim3_nodup.fastq -l CGTGAT:R701 -b 6:6 -s
```

4. Run Trimmomatic in single-ended mode to trim the 5′-end adapter sequences. Use function "HEADCROP," with 16 threads, base quality encoding phred33. Specify input and output files.

```
$ java -jar trimmomatic-0.36.jar SE -threads 16 -phred33 bcnn.fastq x_trim_nodup_bcnn.fastq HEADCROP:17 MINLEN:20
```

3.1.2 FastQC

1. Visually inspect the quality of the preprocessed data using FastQC: "File" > "Open" (*see* Figs. 3 and 4).

3.2 Mapping to Reference Genome, Filtering Reads, Basic Statistics

After data preprocessing, reads can be mapped to a reference genome of choice and classified into different alignment types, filtered for splice junction and short gapped reads, and plotted for gap and segment lengths.

3.2.1 STAR Mapping

1. Initialize STAR mapping to a reference genome with permissive parameters in a script (*see* **Notes 10** and **11**).

```
/STAR-2.7.1a/bin/Linux_x86_64/STAR \
--runThreadN 8 \
--runMode alignReads \
--genomeDir StaridxPath \
--readFilesIn Fastq \
--outFileNamePrefix Outprefix_1_ \
--genomeLoad NoSharedMemory \
--outReadsUnmapped Fastx \
--outFilterMultimapNmax 10 \
--outFilterScoreMinOverLread 0 \
--outSAMattributes All \
--outSAMtype BAM Unsorted SortedByCoordinate \
--alignIntronMin 1 \
--scoreGap 0 \
--scoreGapNoncan 0 \
--scoreGapGCAG 0 \
```

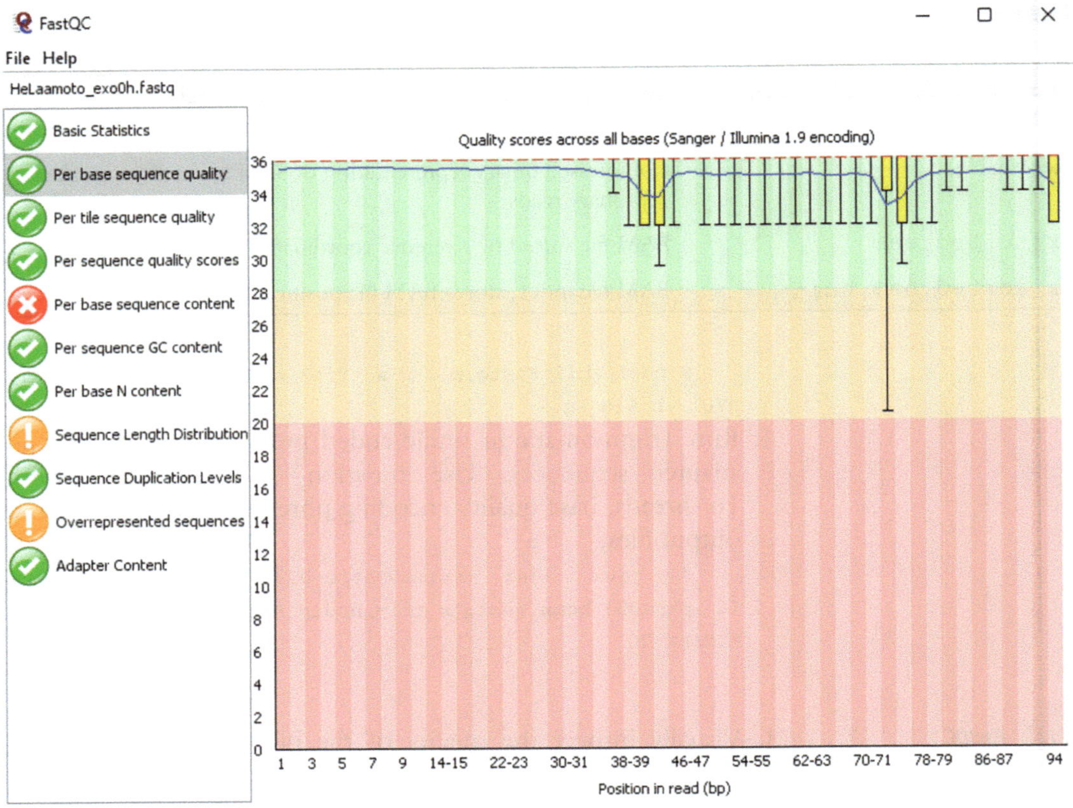

Fig. 3 Screenshot of a fastq file loaded into FastQC. From the "per base sequence quality plot," average quality score across all read positions should be high (≥30)

```
--scoreGapATAC 0 \
--scoreGenomicLengthLog2scale -1 \
--chimOutType WithinBAM HardClip \
--chimSegmentMin 5 \
--chimJunctionOverhangMin 5 \
--chimScoreJunctionNonGTAG 0 \
--chimScoreDropMax 80 \
--chimNonchimScoreDropMin 20
```

Fastq = **x**_trim_nodup_bcnn.fastq

Outprefix = **x**_1_

2. Extract primary alignments using samtools and awk.

```
$ samtools view -h x_Aligned.sortedByCoord.out.bam | awk '$1~/^@/ || NF<21' > x_nonchimeric_temp.sam
$ samtools view -h x_Aligned.sortedByCoord.out.bam | awk '$1!~/^@/ && NF==21'> x_chimeric_temp.sam'
$ samtools view -bS -F 0x900 -o x_nonchimeric_pri.bam x_nonchimeric_temp.sam
$ samtools view -h x_nonchimeric_pri.bam > x_nonchimeric_-pri.sam
```

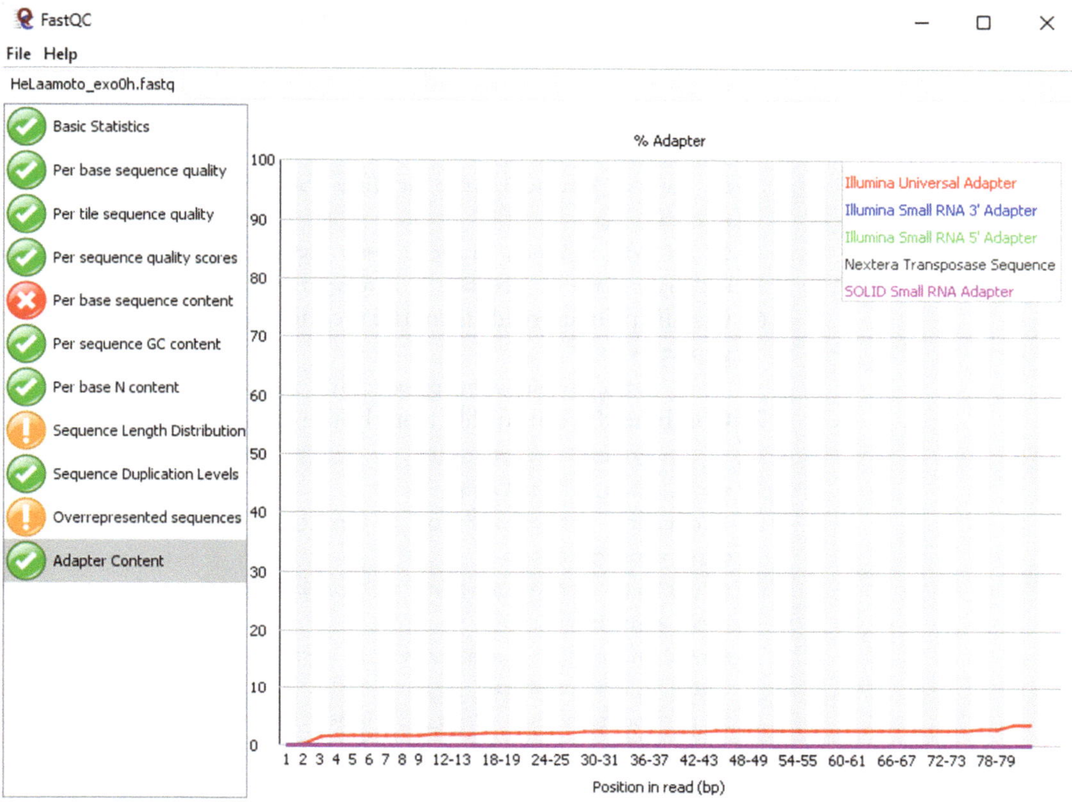

Fig. 4 Screenshot of a fastq file loaded into FastQC. After sequencing adapter trimming steps, adapter content should be close to 0

```
$ cat x_nonchimeric_pri.sam x_chimeric_temp.sam > x_1_pri.
sam
$ rm -f x_nonchimeric_temp.sam x_chimeric_temp.sam x_non-
chimeric_pri.bam x_nonchimeric_pri.sam
```

3. Classify primary reads based on alignment types using gaptypes.py (*see* **Notes 12** and **13**).

```
$ python3 gaptypes.py x_1_pri.sam x_1_pri -1 15 1
```

4. Rearrange any soft-clipped alignments using softreverse.py and map again using STAR with the same parameters as with the above (*see* **Note 14**).

```
$ python softreverse.py x_1_pricont.sam softrev.fastq
/STAR-2.7.1a/bin/Linux_x86_64/STAR \
Fastq = softrev.fastq
Outprefix = x_2_
```

5. Extract primary alignments using samtools and awk again, using different file names.

```
$ samtools view -h x_2_Aligned.sortedByCoord.out.bam |
awk '$1~/^@/ || NF<21' > x_2_nonchimeric_temp.sam
$ samtools view -h x_2_Aligned.sortedByCoord.out.bam | awk
```

```
'$1!~/^@/ && NF==21'> x_2_chimeric_temp.sam'
$ samtools view -bS -F 0x900 -o x_2_nonchimeric_pri.bam x_2_nonchimeric_temp.sam
$ samtools view -h x_2_nonchimeric_pri.bam > x_2_nonchimeric_pri.sam
$ cat x_2_nonchimeric_pri.sam x_2_chimeric_temp.sam > x_2_pri.sam
$ rm -f x_2_nonchimeric_temp.sam x_2_chimeric_temp.sam x_2_nonchimeric_pri.bam x_2_nonchimeric_pri.sam
```

6. Classify primary reads from the second round of mapping using gaptypes.py.

```
$ python3 gaptypes.py x_2_pri.sam x_2_pri -1 15 1
```

7. Combine the output file from both rounds of STAR mapping.

```
$ python3 merger_sams.py x_1_prigap1.sam x_2_prigap1.sam gap1 x_prigap1.tmp
$ python3 merger_sams.py x_1_prigapm.sam x_2_prigapm.sam gapm x_prigapm.tmp
$ python3 merger_sams.py x_1_prihomo.sam x_2_prihomo.sam homo x_prihomo.tmp
$ python3 merger_sams.py x_1_pritrans.sam x_2_pritrans.sam trans x_pritrans.tmp
$ samtools view -H x_1_Aligned.sortedByCoord.out.bam > header
$ cat header x_prigap1.tmp > x_prigap1.sam
$ cat header x_prigapm.tmp > x_prigapm.sam
$ cat header x_prihomo.tmp > x_prihomo.sam
$ cat header x_pritrans.tmp > x_pritrans.sam
$ rm -f header *.tmp
```

3.2.2 Filtering Splices and Short Gaps

8. Remove all splice junction alignments and short deletions (1–2 nt gaps) from "gap1" and "gapm" read files using gapfilter.py.

```
$ python gapfilter.py Gtf x_prigap1.sam x_prigap1_filtered.sam 13 yes
$ python gapfilter.py Gtf x_prigapm.sam x_prigapm_filtered.sam 13 yes
```

3.2.3 Primary Reads Statistics (Segments, Gaps)

1. Plot the length of the gaps for filtered gap1 and gapm reads using gaplendist.py (*see* Fig. 5a, **Notes 15** and **16**).

```
$ cat x_prigap1_filtered.sam x_prigapm_filtered.sam > x_prigaps_filtered.sam
$ python gaplendist.py x_prigaps_filtered.sam sam x.list all
$ python gaplendist.py x.list list x_gaplen.pdf all
$ rm -f x_prigaps_filtered.sam x.list
```

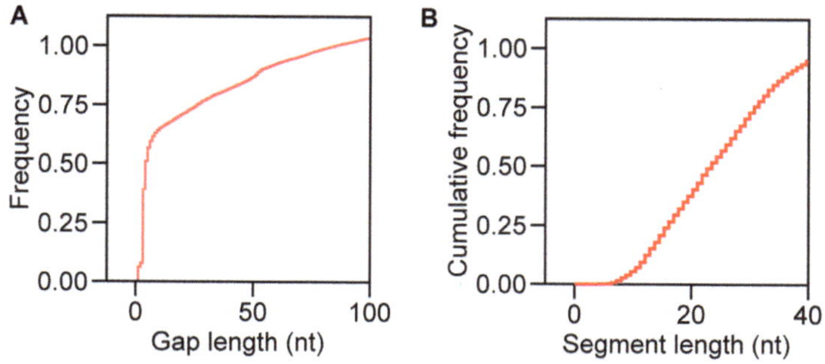

Fig. 5 Example result (a) gaplendist.py, distribution of "gap" lengths in STAR aligned reads (b) seglendist.py, size distribution of "segment" length in STAR aligned reads

2. Plot the length of each segment and distribution for all types of gapped reads using seglendist.py (*see* Fig. 5b, **Notes 16** and **17**).

```
$ cat x_prigap1_filtered.sam x_prigapm_filtered.sam x_pritrans.sam > x_prifiltered.sam
$ python seglendist.py x_prifiltered.sam sam x_prifiltered.list
$ python seglendist.py x_prifiltered.list list x_seglen.pdf
$ rm -f x_prifiltered.sam x_prifiltered.list
```

3. Convert the primary reads file to BAM format. Determine the abundance of mapped reads to specific genes using script CountCdsUtr.py, which yields a summary of quantity and type of primary reads to the CDS, 5′-UTR, and 3′-UTR for genes in the hg38 genome (*see* **Notes 18** and **19**).

```
$ samtools view -bS -o x_prigap1_filtered.bam x_prigap1_filtered.sam
$ python CountCdsUtr.py x_prigap1_filtered.bam CdsUtr.bed none outname
Example: $ python CountCdsUtr.py test_prigap1_filtered.bam CdsUtr.bed none test
```

3.3 CRSSANT—DG Assembly and Analysis

After STAR mapping, separation by gap type, and filtering, reads are output into distinct files as previously described. Depending on the end goal of the analysis, many options are available for further processing. For example, all types of gapped reads can be combined and used as input for CRSSANT analysis (i.e., duplex group (DG) and nonoverlapping group (NG) assembly). Alternatively, reads mapped to a specific chromosome or ribosomal subunit can be extracted for quality control or comparison prior to DG or NG assembly. If performing CRSSANT immediately after mapping, modify relevant variables and paths and run DG assembly as part of the complete pipeline (*see* **Note 20**) (Figs. 6, 7, 8, and 9).

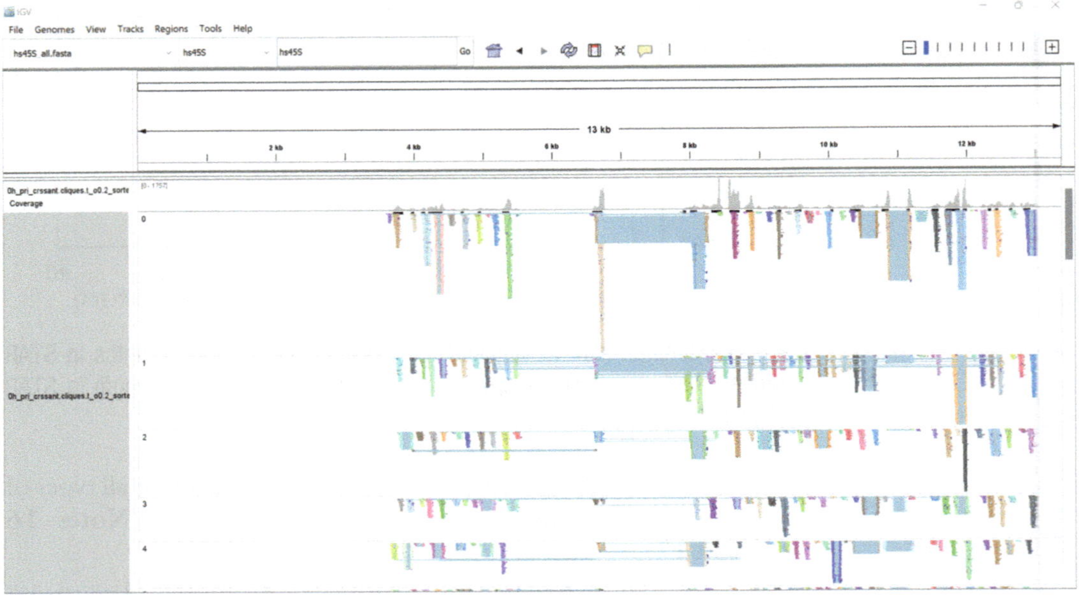

Fig. 6 Screenshot of SAM file loaded into IGV with the hs45S rRNA subunit as reference genome. Visible here are the CRSSANT assembled reads from the hs45S subunit. The track is grouped by nonoverlapping groups (NG), displayed in the "squished" setting

3.3.1 CRSSANT DG Assembly

1. Use script merger.py to combine filtered gap1 and trans SAM files (*see* **Note 21**).

```
$ python merger_v2.py x_prigap1_filtered.sam none x_pri_crssant.sam
```

2. For different sample conditions, tag reads separately with a label according to experimental conditions, then combine the files with the sequencing header (*see* **Notes 22** and **23**).

```
$ awk '$0!~/^@/ {printf $1"-TAG1\t"; for(i=2;i<=NF;++i) printf $i "\t"; printf"\n"}' x_pri_crssant.sam > x_pri_crssant_TAG1_tmp.sam
$ awk '$0!~/^@/ {printf $1"-TAG2\t"; for(i=2;i<=NF;++i) printf $i "\t"; printf"\n"}' x_pri_crssant.sam > x_pri_crssant_TAG2_tmp.sam
$ samtools view -H x_pri_crssant.sam > header
$ cat header x_pri_crssant_TAG1_tmp.sam x_pri_crssant_TAG2_tmp.sam > x_pri_crssant.sam
$ rm -f header x_pri_crssant_TAG1_tmp.sam x_pri_crssant_TAG2_tmp.sam
```

3. Build a bed file for CRSSANT analysis (*see* **Note 24**). This can be done manually by following the tab delimited format described in the CRSSANT manual or by using genes2bed.py to process the "ReadCount.txt" output from CountCdsUtr.py.

```
$ python genes2bed.py x_ReadCount.txt CdsUtr.bed min_reads outname
```

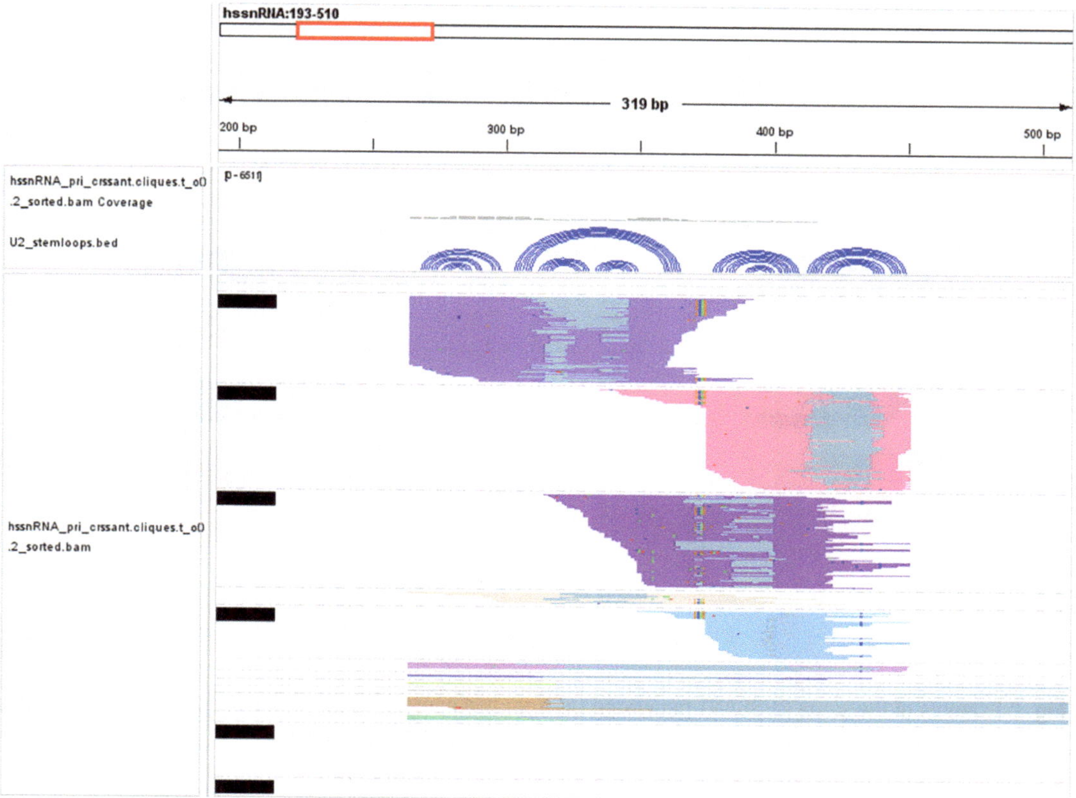

Fig. 7 Example of using *dg2arc.py* to generate an arc file for U2 snRNA to show read arm interactions

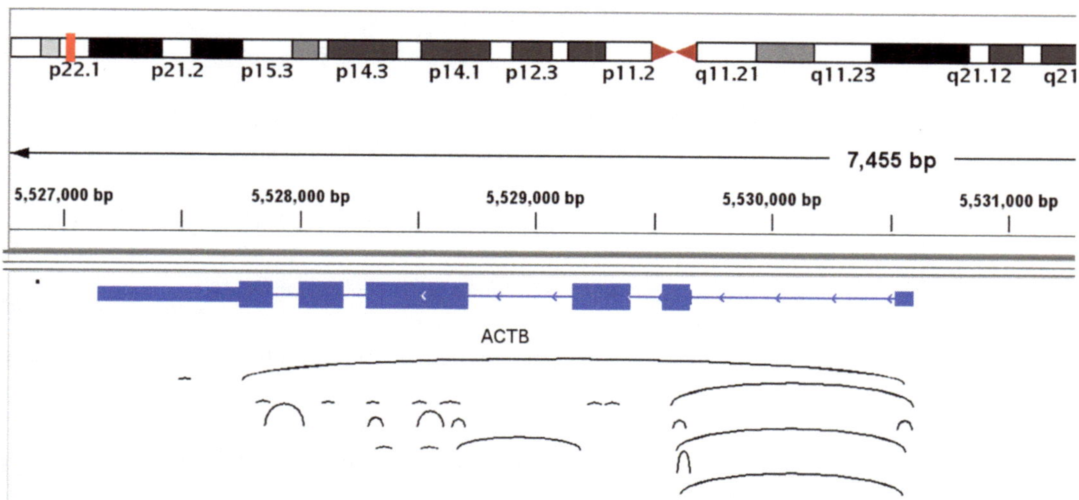

Fig. 8 Example of using bedpetobed12.py to generate arc files for ACTB to show DG interactions

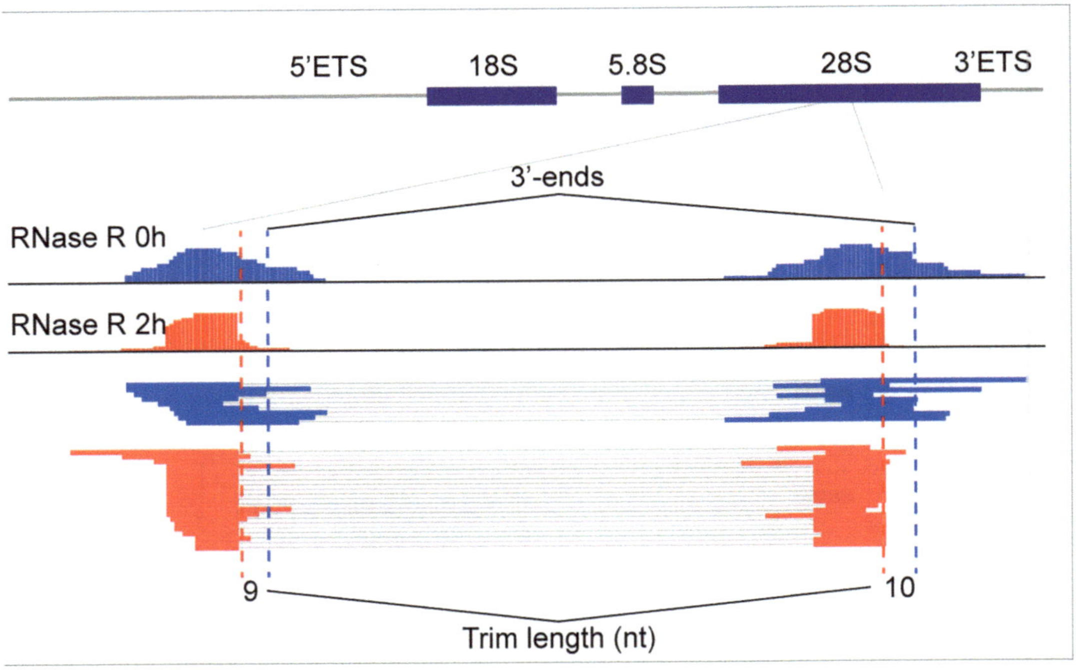

Fig. 9 Example of the same CRSSANT generated DG from two sample types, with endpoints determined. Note the shortened read lengths in the RNase R treated "exo" (exonuclease) sample

```
Example: $ python genes2bed.py test_ReadCount.txt CdsUtr.bed 10 min10
```

4. Use bedtools to generate a bedgraph for CRSSANT analysis.

```
$ bedtools genomecov -bg -split -strand + -ibam x_pri_crssant.bam -g staridxPath/chrNameLength.txt > x_plus.bedgraph
$ bedtools genomecov -bg -split -strand - -ibam x_pri_crssant.bam -g staridxPath/chrNameLength.txt > x_minus.bedgraph
```

5. Initialize CRSSANT for DG and NG assembly (*see* **Note 25**).

```
$ python crssant.py -cluster cliques -t_o 0.2 -out ./ x_pri_crssant.sam Bedx_plus.bedgraph,x_minus.bedgraph
Bed = bed file from step 3.
```

6. Cluster gapm alignments to tri-segment groups (TGs) using the output from the DG assembly.

```
$ python gapmcluster.py x.bedpe x_gapm.sam
```

7. Separate DGs by their tag after CRSSANT assembly is completed.

```
$ awk '$0~/^@/ || $1~/-TAG/' x_pri_crssant.cliques.t_o0.2.sam > x_ crssant_TAG1.sam
```

3.3.2 Ribosomal RNA (rRNA) DG Assembly Example

Since rRNA reads are highly abundant, they are useful for comparative and quality control purposes. However, given the typically large number of rRNA reads, processing every read is unnecessary. Extracting reads mapped to the ribosome and using a randomized subset of them for further analysis is a convenient and recommended approach (Figs. 10 and 11).

8. After mapping, extract only reads mapped to the ribosome and remove the SA tag (*see* **Notes 26** and **27**).

```
$ awk '$0~/^@/ {print}; $0!~/^@/ {if ($0~"hs45S") print}' x_prigap1_filtered.sam > x_prigap1_hs45S.tmp
$ python3 merger_v2.py x_prigap1_hs45S.tmp none x_hs45S.sam
$ rm -f x_prigap1_hs45S.tmp
```

9. Extract a predefined number of randomly selected reads using extract_random.py (*see* **Notes 28** and **29**).

```
$ python extract_random.py x_hs45S.sam @ 30000x_hs45S_30000.sam
```

10. Repeat CRSSANT steps as previously described, this time using a bed file containing "genes" in the hs45S rRNA subunit (*see* **Note 30**).

```
$ samtools view -bS -o x_pri.bam x_hs45S_30000.sam
$ bedtools genomecov -bg -split -strand + -ibam x_hs45S_30000.bam -g staridxPath/chrNameLength.txt > x_plus.bedgraph
$ bedtools genomecov -bg -split -strand - -ibam x_hs45S_30000.bam -g staridxPath/chrNameLength.txt > x_minus.bedgraph
```

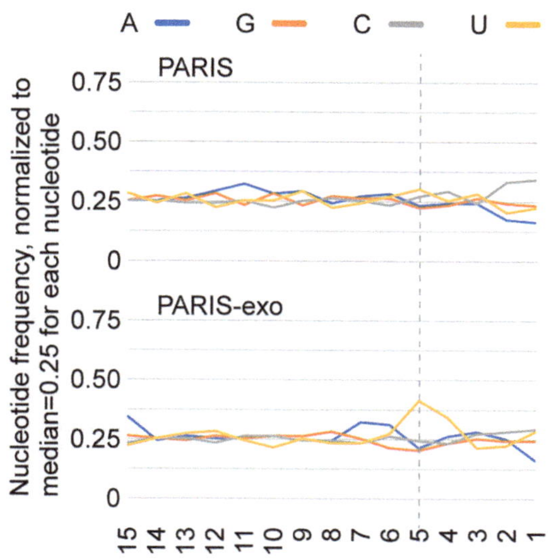

Fig. 10 Example of two sets of data with nucleotide frequency determined for the 18S, 5.8S, and 28S rRNA, 30,000 reads. Note the peak uridine frequency from the "exo" (exonuclease) treated sample

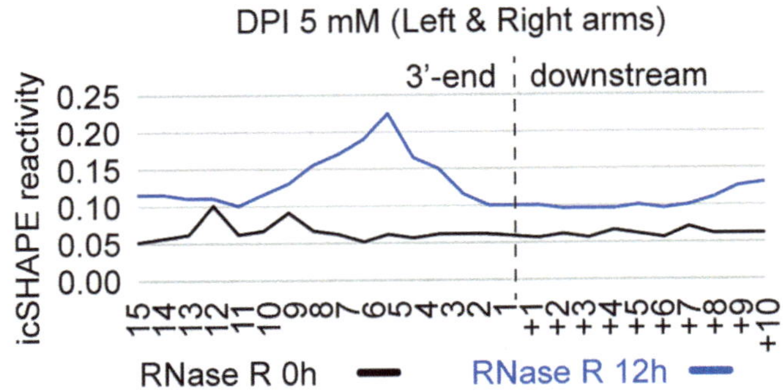

Fig. 11 Example of two sets of data with SHAPE reactivity determined at positions −15 to 10 from the read arm ends. Note the sharp peak in reactivity 5 nucleotides from the arm end in the RNase R treated sample

```
$ python crssant.py -cluster cliques -t_o 0.2 -out ./
x_hs45S_30000.sam Bedx_plus.bedgraph,x_minus.bedgraph
```

3.3.3 DG Analysis

Data produced after duplex group (DG) assembly via CRSSANT can be visualized with various tools.

1. Convert the CRSSANT output SAM file into BAM format, sort, and index. The samtools package is typically used in a three-step process as follows (*see* **Notes 31** and **32**).

```
$ samtools view -bS -o x_pri.bam x_pri.sam
$ samtools sort x_pri.bam x_pri_sorted.bam
$ samtools index x_pri_sorted.bam
```

2. Open the BAM file in IGV, using the appropriate reference genome (*see* **Note 33**). Visualize the DGs, grouping by tag "DG" or "NG."

3. Convert base pairs of DGs to arcs for IGV visualization (*see* **Note 34**).

```
$ python bedpetobed12.py x.cliques.t_o0.2_dg.bedpe x.cliques.t_o0.2_dg.bed
$ sortBed -i x.cliques.t_o0.2_dg.bed > x.cliques.t_o0.2_dg_sorted.bed
Add header line track graphType=arc
```

4. Determine read coverage and DG arm endpoints (*see* **Note 35**).

```
$ python3 dg_endpoint.py input.sam
```

5. Count and plot the frequency of nucleotides at a given position from the 3′-end of reads (*see* **Notes 36** and **37**).

```
$ python nt_freq.py sam ref + - rRNA all_reads reads
Example: $ python nt_freq.py x_hs45S.sam hs45S.fasta 15 10 all n 30000
```

6. Count the frequency of SHAPE nucleotide reactivity at a given position from the 3′-end of reads. Useful for determining the

cross-linking site of acylation-based experiments, efficacy of exonuclease trimming on the arm ends (*see* **Note 38**).

```
$ python SHAPE_freq.py inputsam shape_bedgraph DG_reads_cutoff DGcommon_ratio seqlen extendlen DG/reads outputprefix
Example: $ python SHAPE_freq.py 12h_pri_crssant.cliques.t_o0.2.sam HEK293_icSHAPE_hs45S.bedgraph 10 0.2 15 10 DG 12h_test
```

4 Notes

1. Carefully read manuals of each tool and script. Detailed information on the execution of the scripts and programs, their options, and troubleshooting information is available online or within the script. For rna2d3d scripts, see usage instructions by running the python program with no arguments.
2. Relevant sample data to complete the protocol in its entirety can be downloaded from the original PARIS and CRSSANT references. For troubleshooting purposes, sample files are provided.
3. The pipeline allows for automation via shell scripting after completing any troubleshooting. Example shell scripts are provided, all of which require modification of relevant variables and paths to run.
4. While not explicitly stated in each line of the protocol here, it is recommended that scripts, input, and output files be given absolute directories or paths to avoid potential problems.
5. Each step can, and should, initially be run independently. Depending on the experimental design and method of cDNA library preparation prior to sequencing, preprocessing steps and subsequently the file names may not necessarily be the same or omitted altogether.
6. The file sizes of raw sequencing data and number of barcodes passed through the script will affect the run time and output file sizes. For larger datasets, we highly recommend completing the steps on a supercomputer cluster.
7. For 3′-end trimming, parameters are as follow: seed_mismatches = 3, palindrome_clip_threshold = 20, simple_clip_threshold = 10, window_size = 4, window_minimum_quality = 20, minlen = 18. Ensure that the values are consistent with experimental design to not trim into sequencing reads.
8. After readCollapse, the printout to console will show the total reads processed, the unique number of reads after duplicate

removal, and the unique ratio (unique/total). This is a helpful set of numbers for basic quality control. Values can be kept in a spreadsheet for future reference and troubleshooting.

9. Multiple barcodes can be provided by using two colons "::" between "-l" barcode entries (e.g., CGTGAT:R701::ACATCG:R702). Whenever "-l" is used, barcode length "-b" must be specified. Files will be separated into fastq files based on the provided barcode name. Here, reads with barcode "CGTGAT" will be separated into output R701.fastq.
10. Settings for STAR mapping were previously optimized [41]. Briefly, it includes permissive parameters that enable sensitive and specific mapping of short reads with irregular gaps.
11. The parameters here are part of a shell script that initializes STAR mapping. The reference genome (StaridxPath), preprocessed data (Fastq), and output file prefix (Outprefix) must be specified.
12. Reads will be classified into 5 distinct primary types: cont (continuous, non-gapped), gap1 (1 gap, 2 segments), gapm (>1 gaps, >2 segments), trans (two segments, different strands/chromosomes), or homo (overlapping segments) reads. Maintaining a separate record of the file (i.e., sample origin, experimental conditions, etc.) will mitigate confusion over the details when comparing data, as files will have identical suffixes.
13. The original script "gaptypes.py" found in the CRSSANT repository is only compatible with Python2, so we recommend using gaptypes_v2.py (see rna2d3d repository) which is Python3 ready.
14. Details on the softreverse.py script were previously described [41]. Briefly, it accounts for discriminated backward chimeric alignments by rearranging soft-clipped continuous alignments from the first round of STAR mapping, which are then passed through a second round of STAR mapping.
15. The original script "gaplendist.py" found in the CRSSANT repository uses an older version of matplotlib; gaplendist_v2.py is updated to reflect these changes. However, we recommend using gaplendist_mod.py to simplify the plotting (see the rna2d3d repository).
16. For files not filtered for splice junction alignments or short deletions, file input and output names would not include suffix "_filtered." Subsequent file names would also omit the "filtered" suffix.
17. The original script "seglendist.py" found in the CRSSANT repository uses an older version of matplotlib; seglendist_v2.

py is updated to reflect these changes. However, we recommend using seglendist_mod.py to simplify the plotting (see the rna2d3d repository).

18. CdsUtr.bed is a bed file that specifies gene regions (CDS, intron, 5′-UTR, 3′-UTR) and type (protein coding, lncRNA, miRNA, rRNA, snRNA, etc.). Starting from a GTF file, only the longest isoforms of genes were used to manually prepare this bed file.
19. CountCdsUtr.py outputs various text files containing summarized information on the quantity and types of reads. For example, "_biotype.txt" calculates the percentage of mapped reads, while "_readCount.txt" details the frequency of reads mapped to a gene and its regions. Note that the CountCdsUtr script is only a rough estimate of the read number, as it does not account for all isoforms. The example shown performs the read counting on gap1 reads.
20. The CRSSANT repository has been updated to include a python script that automates the pipeline. See the referenced page for details.
21. Replace argument pritrans.sam with any string for this step of analyses where only prigap1 or prigapm reads are used.
22. Tagging the reads is optional, particularly in cases where there is no separate dataset for comparison. However, it is highly recommended, as the reads will be given an identifier for future reference.
23. Substitute TAG with a short string that will distinguish it from other reads. This is especially important for experiments comparing similar data. For example, samples cross-linked at different concentrations might be given tags "0 M" and "1 M."
24. The bed file here is tab separated with columns indicating chromosome, start and stop location, gene name, coverage for DG assembly, and strand. The script genes2bed.py uses the output "_readCount.txt" from CountCdsUtr to extract information for genes with a minimum read count greater than a given number (≥ "min_reads") as cutoff.
25. One of the most common errors from CRSSANT is caused by the inclusion of the SA tag. To check for the SA tag, use grep or less and search column 21.
26. If necessary (i.e., comparing between a control and experimental sample), tag reads according to the conditions and re-add a sequencing header as previously described.
27. RNA reads were mapped to reference genome starhg38addmask14, which places the ribosomal RNA subunit hs45S into a separate "chromosome" named "hs45S."

28. Here, the number of reads = 30,000. When comparing rRNA of two datasets, the number of reads should be limited by the dataset that has fewer reads. While this value can vary significantly between datasets, a good rule of thumb is to not exceed 5 M reads or use less than 0.01 M reads.
29. The SA tag was previously removed using merger.py. Here, since only gap1 reads are being processed for DG assembly, we must manually remove the tag; otherwise, it will trigger an error when running CRSSANT.
30. This bed file contains the entire hs45S subunit, split into the "genes" 5.8S, 18S, 28S.
31. In this example, all primary reads are converted to BAM format. Oftentimes, converting, sorting, and indexing all mapped reads will take some time. Note that the CRSSANT output SAM file will have a suffix (i.e., "x_pri_crssant.cliques.t_o0.2.sam").
32. The convert-sort-index process can be run line-by-line on a Unix/Linux system, using a shell script (see "samtools_igv.sh" in the rna2d3d repository), or simple python program (see "sam2bamsort.py" in the rna2d3d repository). The converted file can be visualized in software such as the Integrative Genomics Viewer (IGV).
33. BAM index files must be stored in the same directory as the BAM file being opened in IGV.
34. DGs must be on the same chromosome and strand to be converted into a bed12 file prior to visualization on IGV. Script dg2arc.py (see rna2d3d repository) is useful for manually curating arc-type interactions between read arms.
35. The output of "dg_endpoint.py" is a CSV file containing the DG number, coverage (how many reads in that DG), and the 50% average start/stop positions of each DG. By extracting the same information from two sets of relevant data, the DG read arm lengths can be compared. Note that this requires performing CRSSANT DG assembly for both sets of data together, which would be combined after tagging with a unique identifier.
36. In the example here, the nucleotide frequency for x_hs45S.sam is determined using hs45S.fasta as the reference. Figure range is 0 nucleotides downstream or to the "right" of the 3′-end and 15 nucleotides upstream or "left" of the 3′-end. All annotated rRNA "genes" are included, but only a subset of 30,000 reads is used.
37. The output file is a CSV file containing the nucleotide position, and frequency of each nucleotide, normalized to a median of 0.25. This information is especially relevant for determining

the cross-linking site of psoralen- or formaldehyde-based experiments, as well as showing the relative efficiency of exonuclease trimming on read arm ends.

38. The output from "SHAPE_freq.py" displays the relative reactivity of nucleotide positions from upstream and downstream relative to the 3′-end of read arms. This is particularly useful in determining the expected cross-linking in acylation-based experiments and efficacy of exonuclease trimming on the arm ends.

Acknowledgments

The authors thank both the current and former lab members of the Lu lab for providing invaluable input and rigorous testing that have resulted in this protocol. Research in the Lu lab has been supported by startup funds from the University of Southern California, the Pathway to Independence Award from NHGRI (R00HG009662), NIGMS (R35GM143068), USC Research Center for Liver Disease (P30DK48522), Illumina and USC Keck Genomics Platform (KGP), Core Lab Partnership Program, the Norris Comprehensive Cancer Center (P30CA014089), and USC Center for Advanced Research Computing.

References

1. Cooper GM (2000) The molecular composition of cells. Sinauer Associates
2. Alberts B et al (2002) The RNA world and the origins of life, 4th edn. Garland Science, New York
3. Higgs PG, Lehman N (2014) The RNA world: molecular cooperation at the origins of life. Nat Rev Genet 16(1):7–17
4. Robertson MP, Joyce GF (2012) The origins of the RNA world. Cold Spring Harb Perspect Biol 4(5)
5. Cech TR (2012) The RNA worlds in context. Cold Spring Harb Perspect Biol 4(7):a006742
6. Hendrickson DG et al (2016) Widespread RNA binding by chromatin-associated proteins. Genome Biol 17(28):28
7. Caprara MG, Nilsen TW (2023) RNA: versatility in form and function. Nat Struct Biol 7(10):831–833
8. Lu Z, Chang HY (2016) Decoding the RNA structurome. Curr Opin Struct Biol 36:142–148
9. Cech TR, Steitz JA (2014) The noncoding RNA revolution-trashing old rules to forge new ones. Cell 157(1):77–94
10. Serganov A, Nudler E (2013) A decade of riboswitches. Cell 152(1–2):17–24
11. Anastasiadou E, Jacob LS, Slack FJ (2017) Non-coding RNA networks in cancer. Nat Rev Cancer 18(1):5–18
12. Quinn JJ, Chang HY (2016) Unique features of long non-coding RNA biogenesis and function. Nat Rev Genet 17(1):47–62
13. Kaikkonen MU, Adelman K (2018) Emerging roles of non-coding RNA transcription. Trends Biochem Sci 43(9):654–667
14. Sun L et al (2019) RNA structure maps across mammalian cellular compartments. Nat Struct Mol Biol 26(4):322–330
15. Sugimoto Y et al (2015) hiCLIP reveals the in vivo atlas of mRNA secondary structures recognized by Staufen 1. Nature 519(7544): 491–494
16. Ding Y et al (2014) In vivo genome-wide profiling of RNA secondary structure reveals

novel regulatory features. Nature 505(7485): 696–700
17. Ozsolak F, Milos PM (2011) RNA sequencing: advances, challenges and opportunities. Nat Rev Genet 12(2):87
18. Strobel EJ, Yu AM, Lucks JB (2018) High-throughput determination of RNA structures. Nat Rev Genet 19(10):615–634
19. Faruqi AR, Henderson R (2007) Electronic detectors for electron microscopy. Curr Opin Struct Biol 17(5):549–555
20. Ferré-D'Amaré AR, Zhou K, Doudna JA (1998) A general module for RNA crystallization. J Mol Biol 279(3):621–631
21. Zhang H, Keane SC (2019) Advances that facilitate the study of large RNA structure and dynamics by nuclear magnetic resonance spectroscopy. Wiley Interdiscip Rev RNA 10(5): e1541
22. Zhang J, Ferré-D'Amaré AR (2014) New molecular engineering approaches for crystallographic studies of large RNAs. Curr Opin Struct Biol 26:9–15
23. Bai X-C, McMullan G, Scheres SHW (2015) How cryo-EM is revolutionizing structural biology. Trends Biochem Sci 40(1):49–57
24. Das R, Karanicolas J, Baker D (2010) Atomic accuracy in predicting and designing noncanonical RNA structure. Nat Methods 7(4): 291–294
25. Weinreb C et al (2016) 3D RNA and functional interactions from evolutionary couplings: cell. Cell 165(4):963–975
26. Miao Z, Westhof E (2017) RNA structure: advances and assessment of 3D structure prediction. Annu Rev Biophys 46:483–503
27. Ma H et al (2022) Cryo-EM advances in RNA structure determination. Signal Transduct Target Ther 7(1):1–6
28. Lu, Z. and H.Y. Chang, The RNA base-pairing problem and base-pairing solutions. Cold Spring Harb Perspect Biol, 2018. 10(12)
29. Zhang J et al (2022) Advances and opportunities in RNA structure experimental determination and computational modeling. Nat Methods 19(10):1193–1207
30. Wan Y et al (2014) Landscape and variation of RNA secondary structure across the human transcriptome. Nature 505(7485):706–709
31. Mustoe AM et al (2018) Messenger RNA structure regulates translation initiation: a mechanism exploited from bacteria to humans. Biochemistry 57(26):3537–3539
32. Rouskin S et al (2014) Genome-wide probing of RNA structure reveals active unfolding of mRNA structures in vivo. Nature 505(7485): 701–705
33. Spitale RC et al (2013) RNA SHAPE analysis in living cells. Nat Chem Biol 9(1):18–20
34. Spitale RC et al (2015) Structural imprints in vivo decode RNA regulatory mechanisms. Nature 519(7544):486–490
35. Helwak A, Tollervey D (2014) Mapping the miRNA interactome by cross-linking ligation and sequencing of hybrids (CLASH). Nat Protoc 9(3):711–728
36. Ziv O et al (2018) COMRADES determines in vivo RNA structures and interactions. Nat Methods 15(10):785–788
37. Sharma E et al (2016) Global mapping of human RNA-RNA interactions. Mol Cell 62(4):618–626
38. Aw JGA et al (2016) In vivo mapping of eukaryotic RNA Interactomes reveals principles of higher-order organization and regulation. Mol Cell 62(4):603–617
39. Zhang M et al (2021) Optimized photochemistry enables efficient analysis of dynamic RNA structuromes and interactomes in genetic and infectious diseases. Nat Commun 12(1):2344
40. Velema WA, Lu Z (2022) Chemical RNA cross-linking: mechanisms, computational analysis, and biological applications. JACS Au
41. Lu Z et al (2016) RNA duplex map in living cells reveals higher-order transcriptome structure. Cell 165(5):1267–1279
42. Van Damme R et al (2022) Chemical reversible crosslinking enables measurement of RNA 3D distances and alternative conformations in cells. Nat Commun 13(1):911
43. Zhang M et al (2022) Classification and clustering of RNA crosslink-ligation data reveal complex structures and homodimers. Genome Res 32(5):968–985
44. Berger B, Yu YW (2022) Navigating bottlenecks and trade-offs in genomic data analysis. Nat Rev Genet 24:1–16
45. Jiang P et al (2022) Big data in basic and translational cancer research. Nat Rev Cancer 22(11):625–639
46. Kristensen VN et al (2014) Principles and methods of integrative genomic analyses in cancer. Nat Rev Cancer 14(5):299–313
47. Lu Z, Matera AG (2014) Vicinal: a method for the determination of ncRNA ends using chimeric reads from RNA-seq experiments. Nucleic Acids Res 42(9):e79
48. Zhou J et al (2020) IRIS: a method for predicting in vivo RNA secondary structures using PARIS data. Quant Biol 8(4):369–381

49. Lu Z, Matera AG (2014) Developmental analysis of spliceosomal snRNA isoform expression. G3 (Bethesda) 5(1):103–110
50. Hawkins DR, Hon GC, Ren B (2010) Next-generation genomics: an integrative approach. Nat Rev Genet 11(7):476–486
51. Lee WH, Li K, Lu Z (2023) Chemical cross-linking and ligation methods for in vivo analysis of RNA structures and interactions. Methods Enzymol 691:253
52. Lu Z, Gong J, Zhang QC (2018) PARIS: psoralen analysis of RNA interactions and structures with high throughput and resolution. Methods Mol Biol 1649:59–84
53. Bolger AM, Lohse M, Usadel B (2014) Trimmomatic: a flexible trimmer for Illumina sequence data. Bioinformatics (Oxford, England) 30(15):2114–2120
54. Li P, Shi R, Zhang QC (2020) icSHAPE-pipe: a comprehensive toolkit for icSHAPE data analysis and evaluation. Methods (San Diego, Calif) 178:96–103
55. Wingett SW, Andrews S (2018) FastQ screen: a tool for multi-genome mapping and quality control. F1000Research 7:1338
56. Dobin A et al (2013) STAR: ultrafast universal RNA-seq aligner. Bioinformatics (Oxford, England) 29(1):15–21
57. Danecek P et al (2021) Twelve years of SAMtools and BCFtools. GigaScience 10(2): giab008
58. Li H et al (2009) The sequence alignment/map format and SAMtools. Bioinformatics (Oxford, England) 25(16):2078–2079
59. Quinlan AR, Hall IM (2010) BEDTools: a flexible suite of utilities for comparing genomic features. Bioinformatics (Oxford, England) 26(6):841–842
60. Quinlan AR (2014) BEDTools: the Swiss-Army tool for genome feature analysis. Curr Protoc Bioinformatics 47(11.12):1–34
61. Robinson JT et al (2011) Integrative genomics viewer. Nat Biotechnol 29(1):24–26
62. Thorvaldsdóttir H, Robinson JT, Mesirov JP (2023) Integrative genomics viewer (IGV): high-performance genomics data visualization and exploration. Brief Bioinform 14(2): 178–192

Chapter 4

ChRD-PET: A Method for Mapping RNA–DNA Interactions

Qin Xiao and Xingwang Li

Abstract

Chromatin-associated RNAs have important regulatory functions and can promote the formation of spatial compartments in the nucleus. But their conjugated genomic loci and numerous roles in the genome remain largely unknown. Further, we described an efficient method for mapping site-specific chromatin-associated RNA–DNA interactions followed by paired-end-tag sequencing (ChRD-PET), based on chromatin immunoprecipitation and RNA–DNA proximity ligation. A pipeline was also developed for ChRD-PET data analysis and get high-confident RNA–DNA interactions.

Key words Chromatin-associated RNA, RNA, DNA interactions, ChRD-PET

1 Introduction

Eukaryotic chromatin forms a complex and orderly three-dimensional spatial structure through folding and packaging [1]. The precise positions of DNAs, RNAs, and proteins in the nucleus play an important role in maintaining normal functions of cells, such as chromatin structure, chromatin states, and epigenomic modifications [1–4]. Several methods have been developed to capture genomewide RNA–DNA interactions and provide us with numerous resources to explore the functional impacts of chromatin-associated RNAs [5–12]. In addition, since efficient chromatin immunoprecipitation technologies have been developed and well applied in plants [13–18], we are able to efficiently pull-down site-specific chromatin complexes by antibodies of interest. In ChRD-PET, we design a biotinylated double-strand bridge linker, containing a specifically designed forward strand with adenylated 5′ end for ligation to RNA and 3′ end overhanging a nucleotide T for ligation to DNA through A–T base complementary pairing. Reverse transcription of the proximity-ligated RNA is conducted by the extension of the reverse strand of the bridge linker, followed by normal second strand synthesis and library

Michiaki Hamada et al. (eds.), *RNA-RNA and RNA-DNA Interactions*, Methods in Molecular Biology, vol. 2949,
https://doi.org/10.1007/978-1-0716-4670-0_4,

construction using Tn5 transposase. The resulting ChRD-PET library is sequenced, and we developed the ChRD-PET pipeline to analyze ChRD-PET sequence data. (1) Data preprocessing. Adapter sequence and low-quality reads are removed. (2) Extracting DNA and RNA reads. DNA and RNA reads are distinguished according to the direction of the double-strand bridge linker. The upstream sequence connected by the linker represents RNA information, and the downstream sequence represents DNA information in the ChRD-PET reads. (3) Reads alignment. DNA and RNA reads are aligned to the genome respectively. (4) Marking the DNA reads to peak, and the RNA reads to RNA. (5) Getting DNA–RNA pairs. Connect the alignment information of DNA and RNA into DNA–RNA pairs through read ID. (6) Getting high-confidence DNA–RNA interactions. Merging DNA–RNA pairs which have the same peak and RNA, and screening high-confidence DNA–RNA interaction through hypergeometric distribution test [18].

2 Materials

2.1 Sample Cross-Linking

1. Vacuum drier.
2. 1% formaldehyde buffer: Add 4 mL of 37% formaldehyde and144 mL of 1× PBS buffer to a 500-mL beaker, mix the solution well; the formaldehyde fixative solution should be freshly prepared before use.
3. 1.5 mM EGS buffer: Dissolve 0.06 g EGS in 600 μL of DMSO, then add the EGS-DMSO solution to 89.4 mL of 1× PBS buffer; mix the solution well. The EGS fixative solution should be freshly prepared and incubated at 37 °C before use.

2.2 Chromatin Immunoprecipitation

1. Razor blade.
2. Magnetic frame.
3. 1× PBST buffer: Add 44.95 mL of water, 5 mL of 1× PBS buffer and 50 μL of Tween 20 to a 50-mL tube; mix the solution well. The buffer can be stored at 4 °C for up to 2 months.
4. 1% SDS FA buffer: Add 20 mL of water to a 50-mL beaker first, and then add 2.5 mL of HEPES buffer (1 M, pH 7.5), 1.5 mL of NaCl (5 M), 0.1 mL of EDTA (0.5 M), 0.5 mL of Triton X-100, 0.5 mL of sodium deoxycholate (10% (wt/vol)), and 5 mL of SDS (10% (wt/vol)); mix the solution well and bring the final volume to 50 mL. The buffer can be stored at room temperature for up to 6 months.
5. 0.1% SDS FA cell lysis buffer: Add 20 mL of water to a 50-mL beaker first, and then add 2.5 mL of HEPES buffer (1 M, pH 7.5), 1.5 mL of NaCl (5 M), 0.1 mL of EDTA (0.5 M),

0.5 mL of Triton X-100, 0.5 mL of sodium deoxycholate (10% (wt/vol)), and 0.5 mL of SDS (10% (wt/vol)); mix the solution well and bring the final volume to 50 mL. The buffer can be stored at room temperature for up to 6 months.

6. FA buffer without SDS: Add 20 mL of water to a 50-mL beaker first, and then add 2.5 mL of HEPES buffer (1 M, pH 7.5), 1.5 mL of NaCl (5 M), 0.1 mL of EDTA (0.5 M), 0.5 mL of Triton X-100, and 0.5 mL of sodium deoxycholate (10% (wt/vol)), mix the solution well and bring the final volume to 50 mL. The buffer can be stored at room temperature for up to 6 months.
7. High-salt ChIP buffer: Add 32.4 mL of water to a 50-mL beaker first, and then add 2.5 mL of HEPES buffer (1 M, pH 7.5), 3.5 mL of NaCl (5 M), 0.1 mL of EDTA (0.5 M), 0.5 mL of Triton X-100, 0.5 mL of sodium deoxycholate (10% (wt/vol)), and 0.5 mL of SDS (10% (wt/vol)), mix the solution well and bring the final volume to 50 mL. The buffer can be stored at 4 °C for up to 2 months.

2.3 RNA–DNA Proximity Ligation

1. Bridge linker oligos: Forward strand: 5′-[Phos]CCTCGATAC/iBIOdT/TGAGCTGACT-3′; reverse strand: 5′-[Phos]GTCAGC/iBIOdT/CAAGTATCGAG-3′ (*see* **Note 1**).
2. ChRD-PET wash buffer: Add 888 mL of water to a 1-L beaker first, and then add 100 mL of NaCl (5 M), 10 mL of Tris–HCl (1 M, pH 7.5), and 2 mL of EDTA (0.5 M), mix the solution well, and bring the final volume to 1 L. The buffer can be stored at 4 °C for up to 2 months.

2.4 ChRD-PET Library Preparation

1. TruePrep™ DNA Library Prep Kit from Vazyme.
2. iBlock buffer: Dissolve 0.2 g of I-Block Protein-Based Blocking Reagent in 9.5 mL of water in a 65 °C water bath with shaking, and then add 0.5 mL of 10% (wt/vol) SDS to the solution. The buffer can be stored at room temperature for up to 6 months (*see* **Note 2**).
3. 1× binding and wash buffer: Add 39.4 mL of water to a 50-mL beaker first, and then add 10 mL of NaCl (5 M), 0.5 mL of Tris–HCl (1 M, pH 8.0), and 0.1 mL of EDTA (0.5 M); mix the solution well. The buffer can be stored at room temperature for up to 6 months.
4. 2× binding and wash buffer: Add 29.4 mL of water to a 50-mL beaker first, and then add 20 mL of NaCl (5 M), 0.5 mL of Tris–HCl (1 M, pH 8.0), and 0.1 mL of EDTA (0.5 M); mix the solution well. The buffer can be stored at room temperature for up to 6 months.
5. 2× SSC/0.5% SDS buffer: Add 42.5 mL of water to a 50-mL beaker first, and then add 5 mL of 20× SSC and 2.5 mL of SDS

(10% (wt/vol)); mix the solution well. The buffer can be stored at room temperature for up to 3 months.

2.5 Hardware

1. Centos7 operating system with multicore processors. The operations described were performed on a LSF server cluster (Intel(R) Xeon(R) CPU E5-2680 v2 @ 2.80 GHz, 128 GB memory).

2.6 Software

1. Conda, Python3, and packages HTSeq [19], deepTools [20], and Cutadapt [21]. For centos, we chose the miniconda3 download from https://docs.conda.io/en/latest/miniconda.html and installed the conda with default parameters.

   $ wget https://repo.anaconda.com/miniconda/Miniconda3-latest-Linux-x86_64.sh

   $ bash Miniconda3-latest-Linux-x86_64.sh

   $ pip install HTSeq==0.11.2 deepTools==3.2.1 cutadapt

2. SAMtools [22], Bowtie2 [23], HISAT2 [24], bwa [25], flash [26], Trimmomatic [27], BEDtools [28], FastQC, Picard, MACS2 [29], GffRead [30], perl, and r. After conda is installed, these softwares can be installed using the following commands:

   $ conda install samtools bowtie2 hisat2 bwa flash Trimmomatic bedtools fastqc sratools gffread picard macs2

   $ conda install -c conda-forge r-base perl

3 Methods

3.1 Sample Harvesting and Cross-Linking

1. Harvest and clean the plant tissue of interest, then cut into small pieces (*see* **Note 3**).
2. Transfer the sample into 148 mL 1% formaldehyde buffer and vacuum for 15 min with occasionally and gently releasing the gas every 5 min; then, turn off the vacuum (*see* **Note 4**).
3. Add 13.2 mL 2.5 M glycine solution to quench formaldehyde, mix well, and vacuum for 5 min.
4. Rinse the samples three times with deionized water. Discard the supernatant.
5. Merge the samples with 90 mL1.5 mM EGS buffer, mix well, and apply vacuum for 40 min, gently releasing the gas every 10 min; then, turn off the vacuum.
6. Rinse the samples three times with deionized water. Discard the supernatant and remove the surface water with paper towels.
7. Quick-freeze the sample in liquid nitrogen and keep at −80 °C freezer or use immediately.

3.2 Chromatin Immunoprecipitation

1. Take 700 uL of resuspended protein G magnetic beads into the tube and discard the supernatant with the help of a magnetic frame, then wash the beads twice with 1× PBST buffer and resuspend them in 900 uL of 1× PBST buffer with 60 μg antibody you are interested in, incubate and rotate at 4 °C for at least 6 h.
2. Take 1 g sample and ~500 μL 1× PBS into a Petri dish on ice and chop the tissue with a razor blade to release the nuclei. Filter the suspension through a 50-μm mesh filter into a new tube. Spin down at 2800× *g* for 5 min at 4 °C.
3. Resuspend the nuclei pellet with 100 μL 1% SDS FA buffer, vortex, and mix well, and then incubate at 4 °C for 30 min with rotation at 11 RPM.
4. Add 200 μL FA buffer without SDS, mix well, and then incubate at 4 °C for 30 min with rotation at 11 RPM.
5. Place the chromatin solution on ice and add 300 μL FA buffer without SDS, mix and separate into four tubes, remove bubbles by a pipette.
6. Shear the chromatin on a sonication system under specific parameters (Mode: High; Cycs: 28cycs, 30 s ON, 50 s OFF) (*see* **Note 5**).
7. Spin the sonicated chromatin at 12,000× *g* for 10 min at 4 °C, then collect the supernatant and take out 30 μL as input to test the length of the chromatin fragments, keep at −20 °C (*see* **Note 6**); the remaining samples are completed to 2 mL with FA buffer without SDS and proceed to the next step immediately.
8. For chromatin preclearing, take 500 μL resuspended protein G magnetic beads into the tube and discard the supernatant with the help of a magnetic frame, then wash the bead with 1× PBST buffer twice and resuspend the beads with the sample in **step** 7, mix well, and incubate at 4 °C for 2 h with rotation at 11 RPM.
9. Place the tube on a magnetic frame for 2 min and transfer the supernatant to a new tube.
10. Place the tube containing beads and antibody from **step 1** on a magnetic frame for 30 s, discard the supernatant, and wash the beads with 1× PBST buffer twice. Then, resuspend the beads with the supernatant in **step 9**, mix well, and incubate at 4 °C for 8 h with rotation at 11 RPM.
11. Place the tube on a magnetic frame for 30 s, discard supernatant, and wash the beads with the buffer below:

    1 mL 0.1% SDS FA cell lysis buffer, rotate at 4 °C for 5 min and repeat twice.

1 mL high-salt ChIP buffer, rotate at 4 °C for 5 min and repeat once.

1 mL LiCl buffer, rotate at 4 °C for 5 min.

1 mL 1× TE buffer, in situ rotate-washing and repeat once, then proceeded to the next step immediately.

3.3 RNA–DNA Proximity Ligation

1. Resuspend beads in 700 μL of T4 DNA polymerase master mix, containing 600 μL of ddH_2O, 70 μL of 10× buffer for T4 DNA ligase, 7 μL of 10 mM dNTPs, 7 μL of T4 DNA polymerase, 7 μL of T4 PNK, and 9 μL RI. Mix gently and incubate at 37 °C for 40 min with rotation at 11 RPM.
2. Place the tube on a magnetic frame for 30 s, discard the supernatant and wash the beads with ice-cold ChRD-PET wash buffer three times.
3. Resuspend beads in 700 μL of T4 DNA polymerase master mix, containing 610 μL of ddH_2O, 70 μL of 10× NEB Buffer 2, 7 μL of 10 mM dATP, 7 μL of Klenow (3′-5′exo-), and 7 μL RI. Mix gently and incubate at 37 °C for 50 min with rotation at 11 RPM.
4. Place the tube on magnetic frame for 30 s, discard supernatant, and wash the beads with ice-cold ChRD-PET wash buffer three times.
5. For linker preparation, 5 end of forward strand linker is adenylated by Mth RNA ligase enzyme in the reaction mix, containing 51.7 μL of ddH_2O, 10 μL of 10× 5′ DNA Adenylation Reaction Buffer, 9 uL of Mth RNA Ligase enzyme, 9 uL of 100 uM of sense strand linker, 10 uL of 10 mM ATP, and 0.3 uL of 100 mM ATP. Mix gently and incubate at 65 °C for 1 h, 85 °C for 5 min and 4 °C hold. The 9 μL of reverse strand linkers are then annealed to the sense strand linker by incubating the sample at 95 °C for 2 min, followed by cooling down to 22 °C at a rate of 0.1 °C/s. Then purify the oligo and elute in 10 ul of Qigen EB (*see* **Note 1**).
6. Resuspend the beads in **step 4** with 400 μL of RNA ligase mix, containing 158.5 μL of ddH_2O, 40 μL of 10× RNA ligase buffer, 12.5 μL of T4 RNA ligase (Trunked KQ), 160 μL of 50% PEG 8000 (20%), 4 μL of 10% Triton X-100 (0.1%), 20 μL of linker, and 5 μL of RI. Mix gently and incubate at 25 °C for 1 h, 22 °C for 4 h, 16 °C overnight with rotation at 11 RPM.
7. Place the tube on magnetic frame for 1 min, discard supernatant and wash the beads with 37 °C ChRD-PET wash buffer for three times.
8. Resuspend the beads with 200 μL of RNA reverse transcription mix, containing 135 μL of ddH_2O, 40 μL of 5× MMLV buffer, 8 μL of MMLV enzyme (RNase H-), 12 μL of 10 mM dNTP,

and 5 μL of RI. Mix gently and incubate at 37 °C for 40 min with rotation at 11 RPM.

9. Place the tube on a magnetic frame for 1 min and discard the supernatant, then wash the beads with 37 °C ChRD-PET wash buffer three times and ice-cold TE buffer once.
10. Resuspend the beads with 5 mL of T4 DNA ligase mix, containing 3.97 mL ddH_2O, 1 mL of 5× T4 DNA ligase buffer, and 30 μL of T4 DNA ligase enzyme. Mix gently and incubate at 16 °C overnight with rotation at 11 RPM.
11. Place the tube on a magnetic frame for 30 s, discard the supernatant, and wash the beads with ice-cold ChRD-PET wash buffer three times.
12. Resuspend the beads with 400 μL of second strand synthesis mix, containing 295.5 μL of ddH_2O, 80 μL of 5× second strand buffer, 2.7 μL of E. *coli* DNA ligase enzyme, 10.7 μL of DNA polymerase I, 2.7 μL of RNase H, and 8 μL of 10 mM dNTP. Mix gently and incubate at 16 °C for 2 h with rotation at 11 RPM.
13. Add 6 μL of T4 DNA polymerase to the sample, mix gently, and incubate at 16 °C for 5 min.
14. Add 5 μL of RNase A to the sample, mix gently, and incubate at 37 °C for 30 min.
15. Add 10 μL of proteinase K (10 mg/mL) to the sample, mix gently, and incubate at 55 °C for 80 min.
16. Add 15 μL of 5 M NaCl to the sample, mix gently, and incubate at 65 °C overnight.
17. Extract proximity ligation RNA–DNA (plRD) from the sample with equivalent volume of phenol: chloroform: IAA (ph 7.9), mix well, and centrifuge the sample at 13,000 rpm for 5 min.
18. Transfer the upper aqueous phase into a new tube. Then add equivalent volume of isopropanol, 1/10 volume 3 M sodium actetate and 1 μL GlycoBlue (50 μg/mL), mix gently, and incubate at −80 °C for 30 min.
19. Centrifuge the sample at 13,000 rpm for 20 min at 4 °C and carefully wash the pellet with 75% ice-cold ethanol. Make sure the pellet remains in the tube, air-dry the pellet, and dissolve plRD in 21 μL of QIAGEN EB buffer.

3.4 ChRD-PET Library Construction

1. Take 50 ng of the plRD sample to a new tube and add ddH_2O to 50 μL. Digest the sample with the Tn5 enzyme mix, containing 10 μL of 5× TTBL, 5 μL TTE Mix, and 50 μL sample (50 ng plRD). Mix gently and incubate at 55 °C for 10 min. Then, purify the digested plRD with AMPure® XP beads and elute in 50 uL Qiagen EB (*see* **Note 7**).

2. Take out 15 μL Dynabeads M-280 streptavidin into the tube and discard the supernatant with the help of a magnetic frame, then wash the bead with 200 μL 2× binding and wash buffer twice.
3. Resuspend the M-280 beads with 100 μL I-Block buffer and incubate at room temperature with rotation at a specific parameter (UU45, 45 min) on Intelli-mixer RM-2 L. Then wash the beads with 200 μL 1× binding and wash buffer twice.
4. Resuspend the M-280 beads with 50 μL 2× binding and wash buffer and 50 μL sheared genomic DNA (500 ng fragment genomic DNA from other species). Mix gently and incubate at room temperature with rotation at a specific parameter (UU45, 30 min) on Intelli-mixer RM-2 L. Then wash the beads with 200 μL 1× binding and wash buffer twice.
5. Resuspend the M-280 beads with 50 μL of 2× binding and wash buffer and 50 μL of digested plRD in **step 1**. Mix gently and incubate at room temperature with rotation at a specific parameter (UU45, 45 min) on Intelli-mixer RM-2 L. Then wash the beads with 200 μL of 2× SSC/0.5% SDS buffer five times and 200 μL of 1× binding and wash buffer twice.
6. Resuspend the beads with PCR mix, containing 24 μL of ddH_2O, 10 μL of 5× TAB, 1 μL of TAE, 5 μL of PPM, 5 μL of N5XX, and 5 μL of N7XX. Mix gently.
7. Put the PCR tube in the PCR instrument, then set the procedures as follows (*see* **Note 8**):

Heat	On, 105 °C	
72 °C	3 min	
98 °C	30 s	
98°C 60°C	15 30	9 cycles
72 °C	3 min	
72 °C	5 min	
4 °C	Hold	

8. Add 0.55 volume of AMPure® XP Beads to the PCR products, mix well, and incubate at room temperature for 5 min.
9. Place the sample on a magnetic frame until the solution is clarified and transfer the supernatant to a new tube (*see* **Note 9**).
10. Add 0.2 volumes of AMPure® XP beads to the sample in **step 9**, mix well, and incubate at room temperature for 5 min.

11. Place the sample on a magnetic frame until the solution is clarified and carefully remove the supernatant.
12. Keep the tube in the magnetic frame and wash the beads with fresh prepared 80% of ethanol, and carefully remove the supernatant.
13. Repeat the **step 12** and rinse totally twice (*see* **Note 10**).
14. Keep the tube in the magnetic frame and air-dry the beads (*see* **Note 11**).
15. Take out the tube from the magnetic frame and add 16 μL of ddH_2O to resuspend the beads; mix well and incubate at room temperature for 5 min.
16. Place the tube on magnetic frame until the solution is clarified and carefully transfer the supernatant to a new tube. Label the sample as ChRD-PET library.
17. Sequence the ChRD-PET library.

3.5 Genome Preparation

1. Prepare genome files. Download the MH63 genome sequence and annotation file from http://rice.hzau.edu.cn/rice_rs1 and download mitochondrion (https://www.ncbi.nlm.nih.gov/nuccore/109156581?tdsourcetag=s_pctim_aiomsg) and chloroplast (https://www.ncbi.nlm.nih.gov/nuccore/89280701) from NCBI.

   $ wget http://rice.hzau.edu.cn/rice_rs1/download_ext/MH63RS1.LNNK00000000.fsa.tar.gz

   $ wget http://rice.hzau.edu.cn/rice_rs1/download_ext/MH63_chr.gff.tar.gz

   $ wget http://rice.hzau.edu.cn/rice_rs1/download_ext/MH63_repeat.gff3.tar.gz

   $ tar zxf MH63RS1.LNNK00000000.fsa.tar.gz && tar zxf MH63_chr.gff.tar.gz && tar zxf MH63_repeat.gff3.tar.gz

   $ cat MH63RS1.LNNK00000000.fsa mitochondrion.fasta chloroplast.fasta > genome.fa

   $ cat MH63_chr_v2.gff mitochondrion.gff chloroplast.gff > genome.gff3

   $ perl -lane 'if($F[2] eq "gene"){$tmp=(split(/;/,$F[8]))[0];$gene=(split(/=/,$tmp))[1];print "$F[0]\t".($F[3]-1)."\t$F[4]\t$gene\t.\t$F[6]"}' genome.gff3 | awk '{print $0"\t"$4"\tcoding"}' > genome.info

   $ grep -i rrna MH63_repeat.gff3 | awk '{print $1"\t"$4-1"\t"$5}' > genome.rRNA.bed

2. Prepare genome indexes. Convert the annotation file in "gff" format into "gtf" format by GffRead. Generate genome bwa index, HISAT2 index, and Bowtie2 index.

   $ bwa index genome.fa

   $ gffread genome.gff3 -T -o genome.gtf

$ python hisat2/extract_splice_sites.py genome.gtf > genome.ss

$ python hisat2/extract_exons.py genome.gtf > genome.exon

$ hisat2-build -p 10 --ss genome.ss --exon genome.exon genome.fa genome.hisat

$ bowtie2-build --threads 10 genome.fa genome.bowtie2

3.6 ChIP-PET Data Analysis

1. Download MH63 H3K4me3 and input ChIP-Seq data by sratools.

   $ prefetch SRR10751519

   $ prefetch SRR10751520-

   $ prefetch SRR10751667

   $ fastq-dump --split-3 -gzip SRR10751519.sra

   $ fastq-dump --split-3 -gzip SRR10751520.sra

   $ fastq-dump --split-3 -gzip SRR10751667.sra

2. Evaluate the ChIP-Seq data quality with FastQC and remove low-quality reads by Trimmomatic.

   $ fastqc -t 10 SRR10751519_1.fastq.gz SRR10751519_2.fastq.gz SRR10751520_1.fastq.gz SRR10751520_2.fastq.gz SRR10751667_1.fastq.gz SRR10751667_2.fastq.gz

   $ trimmomatic PE -threads 10 -phred33 SRR10751519_1.fastq.gz SRR10751519_2.fastq.gz SRR10751519_R1.trim.fq.gz SRR10751519_R1.un.fq.gz SRR10751519_R2.trim.fq.gz SRR10751519_R2.un.fq.gz HEADCROP:3 LEADING:3 TRAILING:3 AVGQUAL:20 SLIDINGWINDOW:4:15 MINLEN:75

   $ trimmomatic PE -threads 10 -phred33 SRR10751520_1.fastq.gz SRR10751520_2.fastq.gz SRR10751520_R1.trim.fq.gz SRR10751520_R1.un.fq.gz SRR10751520_R2.trim.fq.gz SRR10751520_R2.un.fq.gz HEADCROP:3 LEADING:3 TRAILING:3 AVGQUAL:20 SLIDINGWINDOW:4:15 MINLEN:75

   $ trimmomatic PE -threads 10 -phred33 SRR10751667_1.fastq.gz SRR10751667_2.fastq.gz SRR10751667_R1.trim.fq.gz SRR10751667_R1.un.fq.gz SRR10751667_R2.trim.fq.gz SRR10751667_R2.un.fq.gz HEADCROP:3 LEADING:3 TRAILING:3 AVGQUAL:20 SLIDINGWINDOW:4:15 MINLEN:75

3. ChIP-Seq data is aligned to genome through bwa with a default parameter; remove PCR reads by picard and filter low-mapping quality reads by SAMtools.

   $ bwa mem -t genome.fa SRR10751519_R1.trim.fq.gz SRR10751519_R2.trim.fq.gz | samtools view -Sb - | samtools sort -@ 10 -o SRR10751519.bam –

$ bwa mem -t genome.fa SRR10751520_R1.trim.fq.gz SRR10751520_R2.trim.fq.gz | samtools view -Sb - | samtools sort -@ 10 -o SRR10751520.bam –

$ bwa mem -t genome.fa SRR10751667_R1.trim.fq.gz SRR10751667_R2.trim.fq.gz | samtools view -Sb - | samtools sort -@ 10 -o SRR10751667.bam -

$ java -jar picard.jar MarkDuplicates I=SRR10751519.bam O=SRR10751519.rmd.bam M=SRR10751519.metrics CREATE_INDEX=true VALIDATION_STRINGENCY=SILENT REMOVE_DUPLICATES=false && samtools view -q 20 -F 1024 -o SRR10751519.flt.bam SRR10751519.rmd.bam

$ java -jar picard.jar MarkDuplicates I=SRR10751520.bam O=SRR10751520.rmd.bam M=SRR10751520.metrics CREATE_INDEX=true VALIDATION_STRINGENCY=SILENT REMOVE_DUPLICATES=false && samtools view -q 20 -F 1024 -o SRR10751520.flt.bam SRR10751520.rmd.bam

$ java -jar picard.jar MarkDuplicates I=SRR10751667.bam O=SRR10751667.rmd.bam M=SRR10751667.metrics CREATE_INDEX=true VALIDATION_STRINGENCY=SILENT REMOVE_DUPLICATES=false && samtools view -q 20 -F 1024 -o SRR10751667.flt.bam SRR10751667.rmd.bam

4. Peak calling by MACS2. The peak regions are used as the ChRD-PET analysis input.

   $ samtools merge MH63.H3K4me3.bam SRR10751519.flt.bam SRR10751520.flt.bam

   $ macs2 callpeak -t MH63.H3K4me3.bam -c SRR10751667.flt.bam -f BAM -g 360545087 -n MH63.H3K4me3.narrow --trackline -B --verbose 3 –SPMR

   $ awk -v OFS="\t" '{len=int(($3-$2)*0.1);start=$2-len; if(start<0){start=0;}print $1,start,$3+len,"peak"NR}' MH63.H3K4me3.narrow_peaks.narrowPeak | bedtools sort -i - | awk -v OFS="\t" '{print $1,$2,$3,"anchor"NR}' > MH63.H3K4me3.exd.peak

3.7 ChRD-PET Data Analysis

1. Download ChRD-PET data and evaluate the ChRD-PET sequencing data quality using FastQC.

   $ prefetch SRR13248068

   $ prefetch SRR13248069

   $ fastq-dump --split-3 -gzip SRR13248068.sra

   $ fastq-dump --split-3 -gzip SRR13248069.sra

   $ fastqc -t 10 SRR13248068_1.fastq.gz SRR13248068_2.fastq.gz SRR13248069_1.fastq.gz SRR13248069_2.fastq.gz

2. Remove the sequencing connectors (Tn5 adaptor sequence) by Cutadapt (*see* **Note 12**).

$ trimmomatic PE -threads 10 -phred33 SRR13248068_1.fastq.gz SRR13248068_2.fastq.gz SRR13248068_R1.trim.fq.gz SRR13248068_R1.un.fq.gz SRR13248068_R2.trim.fq.gz SRR13248068_R2.un.fq.gz ILLUMINACLIP: NexteraPE-PE.fa:2:30:7:8:true LEADING:10 TRAILING:10 SLIDINGWINDOW:4:15 MINLEN:30

$ trimmomatic PE -threads 10 -phred33 SRR13248069_1.fastq.gz SRR13248069_2.fastq.gz SRR13248069_R1.trim.fq.gz SRR13248069_R1.un.fq.gz SRR13248069_R2.trim.fq.gz SRR13248069_R2.un.fq.gz ILLUMINACLIP: NexteraPE-PE.fa:2:30:7:8:true LEADING:10 TRAILING:10 SLIDINGWINDOW:4:15 MINLEN:30

3. Extend the length of short reads by overlapping paired-end reads from fragment libraries that are sufficiently short through FLASH.

   $ flash --threads 10 -M 145 -O --output-prefix SRR13248068.flash --output-directory ./ SRR132480681.trim.R1.fq.gz SRR13248068.trim.R2.fq.gz

   $ flash --threads 10 -M 145 -O --output-prefix SRR13248069.flash --output-directory ./ SRR132480691.trim.R1.fq.gz SRR13248069.trim.R2.fq.gz

4. Find the location of the linker by Cutadapt from three files in **step 3**.

   $ echo -e ">A\nCCTCGATACTTGAGCTGACT\n > A_\nAGTCAGCTCAAGTATCGAGG\n" > linker.fa

   $ cutadapt -b file: linker.fa -n 14 --no-indels -o SRR13248068.combine.noLinker --info-file SRR13248068.combine.Linker_info --discard -O 18 SRR13248068.flash.extendedFrags.fastq > SRR13248068.combine.stat

   $ perl ts_cutadapt2oneline.pl SRR13248068.combine.Linker_info SRR13248068.combine.cut.info > SRR13248068.combine.cut.info.log

   $ cutadapt -b file: linker.fa -n 8 --no-indels -o SRR13248068.notCombined.1.noLinker --info-file SRR13248068.notCombined.1.Linker_info --discard -O 18 SRR13248068.flash.notCombined_1.fastq > SRR13248068.notCombined.1.stat

   $ perl ts_cutadapt2oneline.pl SRR13248068.notCombined.1.Linker_info SRR13248068.notCombined.1.cut.info > SRR13248068.notCombined.1.cut.info.log

   $ cutadapt -b file: linker.fa -n 8 —no-indels -o SRR13248068.notCombined.2.noLinker --info-file SRR13248068.notCombined.2.Linker_info --discard -O 18 SRR13248068.flash.notCombined_2.fastq > SRR13248068.notCombined.2.stat

$ perl ts_cutadapt2oneline.pl SRR13248068.notCombined.2.Linker_info SRR13248068.notCombined.2.cut.info > SRR13248068.notCombined.2.cut.info.log

$ cutadapt -b file: linker.fa -n 14 --no-indels -o SRR13248069.combine.noLinker —info-file SRR13248069.combine.Linker_info --discard -O 18 SRR13248069.flash.extendedFrags.fastq > SRR13248069.combine.stat

$ perl ts_cutadapt2oneline.pl SRR13248069.combine.Linker_info SRR13248069.combine.cut.info > SRR13248069.combine.cut.info.log

$ cutadapt -b file: linker.fa -n 8 --no-indels -o SRR13248069.notCombined.1.noLinker --info-file SRR13248069.notCombined.1.Linker_info --discard -O 18 SRR13248069.flash.notCombined_1.fastq > SRR13248069.notCombined.1.stat

$ perl ts_cutadapt2oneline.pl SRR13248069.notCombined.1.Linker_info SRR13248069.notCombined.1.cut.info > SRR13248069.notCombined.1.cut.info.log

$ cutadapt -b file: linker.fa -n 8 --no-indels -o SRR13248069.notCombined.2.noLinker --info-file SRR13248069.notCombined.2.Linker_info --discard -O 18 SRR13248069.flash.notCombined_2.fastq > SRR13248069.notCombined.2.stat

$ perl ts_cutadapt2oneline.pl SRR13248069.notCombined.2.Linker_info SRR13248069.notCombined.2.cut.info > SRR13248069.notCombined.2.cut.info.log

5. Extract DNA and RNA reads from three files in **step 4** (*see* **Note 13**).

$ perl split.combine.pl SRR13248068.combine.cut.info SRR13248068 1 > SRR13248068.combine.fq.stat 2 > SRR13248068.combine.fq.stat.log

$ perl split.only.pl SRR13248068.notCombined.1.cut.info SRR13248068.notCombined.2.cut.info SRR13248068.only 1 > SRR13248068.only.fq.stat 2 > SRR13248068.only.fq.stat.log

$ cat SRR13248068.combine.RNA.fq SRR13248068.only.RNA.fq > SRR13248068.RNA.fastq

$ cat SRR13248068.combine.DNA.fq SRR13248068.only.DNA.fq > SRR13248068.DNA.fastq

$ perl split.combine.pl SRR13248069.combine.cut.info SRR13248069 1 > SRR13248069.combine.fq.stat 2 > SRR13248069.combine.fq.stat.log

$ perl split.only.pl SRR13248069.notCombined.1.cut.info SRR13248069.notCombined.2.cut.info SRR13248069.only 1 > SRR13248069.only.fq.stat 2 > SRR13248069.only.fq.stat.log

$ cat SRR13248069.combine.RNA.fq SRR13248069.only.RNA.fq > SRR13248069.RNA.fastq

$ cat SRR13248069.combine.DNA.fq SRR13248069.only.DNA.fq > SRR13248069.DNA.fastq

6. DNA reads are aligned to MH63 genome by bwa. Unmapped reads and multi-mapped reads are dropped.

   $ perl split.pl SRR13248068.DNA.fastq SRR13248068 70

   $ bwa aln -t 10 -f SRR13248068.DNA.short.sai genome.fa SRR13248068.DNA.short.fastq

   $ bwa samse -f SRR13248068.DNA.short.sam genome.fa SRR13248068.DNA.short.sai SRR13248068.DNA.short.fastq

   $ bwa mem -t 10 genome.fa SRR13248068.DNA.long.fastq > SRR13248068.DNA.long.sam

   $ java -cp LGL.jar LGL.util.UniqueSam SRR13248068.DNA.long.sam SRR13248068.DNA.long.rmdup.sam

   $ python combine_DNA_file.py SRR13248068.DNA.short.sam SRR13248068.DNA.long.rmdup.sam SRR13248068.DNA.sam

   $ samtools view -Sb SRR13248068.DNA.sam | samtools sort -@ $NTHREADS -o SRR13248068.DNA.bam -

   $ samtools index SRR13248068.DNA.bam

   $ perl split.pl SRR13248069.DNA.fastq SRR13248069 70

   $ bwa aln -t 10 -f SRR13248069.DNA.short.sai genome.fa SRR13248069.DNA.short.fastq

   $ bwa samse -f SRR13248069.DNA.short.sam genome.fa SRR13248069.DNA.short.sai SRR13248069.DNA.short.fastq

   $ bwa mem -t 10 genome.fa SRR13248069.DNA.long.fastq > SRR13248069.DNA.long.sam

   $ java -cp LGL.jar LGL.util.UniqueSam SRR13248069.DNA.long.sam SRR13248069.DNA.long.rmdup.sam

   $ python combine_DNA_file.py SRR13248069.DNA.short.sam SRR13248069.DNA.long.rmdup.sam SRR13248069.DNA.sam

   $ samtools view -Sb SRR13248069.DNA.sam | samtools sort -@ $NTHREADS -o SRR13248069.DNA.bam -

   $ samtools index SRR13248069.DNA.bam

7. Firstly, RNA reads are aligned to the MH63 genome by HISAT2. Then, unmapped reads and multi-mapped reads are realigned to the MH63 genome by Bowtie2 with a local model. Finally, unmapped and multi-mapped reads are dropped (*see* **Note 14**).

$ hisat2 —known-splicesite-infile genome.ss -p 10 -x genome.hisat —rna-strandness F -U SRR13248068.RNA.fastq -S SRR13248068.RNA.sam

$ python flt_bam.py SRR13248068.RNA.sam SRR13248068.RNA

$ samtools sort -@ 10 -o SRR13248068.RNA.hisat.bam SRR13248068.RNA.uniqmap.bam

$ samtools view -Sb -F 256 SRR13248068.RNA.multimap.bam | bamToFastq -i - -fq SRR13248068.tmp.multi.fq

$ bamToFastq -i SRR13248068.RNA.unmap.bam -fq SRR13248068.tmp.unmap.fq

$ cat SRR13248068.tmp.multi.fq SRR13248068.tmp.unmap.fq > SRR13248068.remap.fq

$ bowtie2 -p 10 --local --very-sensitive-local -x genome.bowtie2 -U SRR13248068.remap.fq -S SRR13248068.remap.sam

$ samtools view -Sb -q 2 SRR13248068.remap.sam | samtools sort -@ 10 -o SRR13248068.remap.bam -

$ samtools merge -f SRR13248068.RNA.bam SRR13248068.RNA.hisat.bam SRR13248068.remap.bam

$ samtools index SRR13248068.RNA.bam

$ hisat2 —known-splicesite-infile genome.ss -p 10 -x genome.hisat --rna-strandness F -U SRR13248069.RNA.fastq -S SRR13248069.RNA.sam

$ python flt_bam.py SRR13248069.RNA.sam SRR13248069.RNA

$ samtools sort -@ 10 -o SRR13248069.RNA.hisat.bam SRR13248069.RNA.uniqmap.bam

$ samtools view -Sb -F 256 SRR13248069.RNA.multimap.bam | bamToFastq -i - -fq SRR13248069.tmp.multi.fq

$ bamToFastq -i SRR13248069.RNA.unmap.bam -fq SRR13248069.tmp.unmap.fq

$ cat SRR13248069.tmp.multi.fq SRR13248069.tmp.unmap.fq > SRR13248069.remap.fq

$ bowtie2 -p 10 --local --very-sensitive-local -x genome.bowtie2 -U SRR13248069.remap.fq -S SRR13248069.remap.sam

$ samtools view -Sb -q 2 SRR13248069.remap.sam | samtools sort -@ 10 -o SRR13248069.remap.bam -

$ samtools merge -f SRR13248069.RNA.bam SRR13248069.RNA.hisat.bam SRR13248069.remap.bam

$ samtools index SRR13248069.RNA.bam

8. Reads overlapped with the rRNA region are filtered by BEDTools.

   $ mv SRR13248068.DNA.bam SRR13248068.DNA.old.bam && intersectBed -a SRR13248068.DNA.old.bam -b genome.rRNA.bed -v > SRR13248068.DNA.bam

```
$ mv SRR13248068.RNA.bam SRR13248068.RNA.old.bam && intersectBed -a SRR13248068.RNA.old.bam -b genome.rRNA.bed -v > SRR13248068.RNA.bam
$ mv SRR13248069.DNA.bam SRR13248069.DNA.old.bam && intersectBed -a SRR13248069.DNA.old.bam -b genome.rRNA.bed -v > SRR13248069.DNA.bam
$ mv SRR13248069.RNA.bam SRR13248069.RNA.old.bam && intersectBed -a SRR13248069.RNA.old.bam -b genome.rRNA.bed -v > SRR13248069.RNA.bam
```

9. According to H3K4me3 peak region, each DNA read will be assigned to a unique peak by BEDTools. According to gene annotation, each RNA read will be assigned to RNA by HTSeq.

```
$ bamToBed -cigar -i SRR13248068.DNA.bam | awk -v OFS="\t" ’{if($6=="+"){print $1,$2,$2+500,$4,$5,$6,$7} else{start=$3-500;if(start<0){start=0;}print $1,start,$3,$4,$5,$6,$7}}’ | intersectBed -a - -b MH63.H3K4me3.exd.peak -wao | python uniq_DNA.py > SRR13248068.DNA.bed
$ htseq-count -o SRR13248068.RNA.yes.gene.sam -f bam -r name -s yes -t gene -i ID -m intersection-nonempty SRR13248068.RNA.bam genome.gff3 > SRR13248068.RNA.yes.gene.count
$ samtools view -H SRR13248068.RNA.bam | cat - SRR13248068.RNA.yes.gene.sam > SRR13248068.RNA.yes.gene.cg.sam
$ htseq-count -o SRR13248068.RNA.yes.exon.sam -f sam -r name -s yes -t exon -i ID -m intersection-nonempty SRR13248068.RNA.yes.gene.cg.sam genome.gff3 > SRR13248068.RNA.yes.exon.count
$ samtools view -H SRR13248068.RNA.bam | cat - SRR13248068.RNA.yes.exon.sam | perl -lane ’if(/XF:Z/){$_=~s/XF:Z/GE:Z/;print $_}else{print $_}’ > SRR13248068.RNA.yes.sam
$ python sam2bed2.py SRR13248068.RNA.yes.sam 1> SRR13248068.RNA.yes.bed 2> SRR13248068.RNA.yes.bed.log
$ bamToBed -cigar -i SRR13248069.DNA.bam | awk -v OFS="\t" ’{if($6=="+"){print $1,$2,$2+500,$4,$5,$6,$7} else{start=$3-500;if(start<0){start=0;}print $1,start,$3,$4,$5,$6,$7}}’ | intersectBed -a - -b MH63.H3K4me3.exd.peak -wao | python uniq_DNA.py > SRR13248069.DNA.bed
$ htseq-count -o SRR13248069.RNA.yes.gene.sam -f bam -r name -s yes -t gene -i ID -m intersection-nonempty SRR13248069.RNA.bam genome.gff3 > SRR13248069.RNA.yes.gene.count
$ samtools view -H SRR13248069.RNA.bam | cat - SRR13248069.RNA.yes.gene.sam > SRR13248069.RNA.yes.gene.cg.sam
```

$ htseq-count -o SRR13248069.RNA.yes.exon.sam -f sam -r name -s yes -t exon -i ID -m intersection-nonempty SRR13248069.RNA.yes.gene.cg.sam genome.gff3 > SRR13248069.RNA.yes.exon.count

$ samtools view -H SRR13248069.RNA.bam | cat - SRR13248069.RNA.yes.exon.sam | perl -lane 'if(/XF:Z/){$_=~s/XF:Z/GE:Z/;print $_}else{print $_}' > SRR13248069.RNA.yes.sam

$ python sam2bed2.py SRR13248069.RNA.yes.sam 1> SRR13248069.RNA.yes.bed 2> SRR13248069.RNA.yes.bed.log

10. Through the DNA and RNA reads id, merge the DNA and RNA to get the DNA–RNA pair. Then, DNA–RNA pair with MAPQ<20 are removed.

    $ python combineDNAandRNA_withanchor.py SRR13248068.DNA.bed SRR13248068.RNA.yes.bed SRR13248068.DNARNA.bedpe

    $ python combineDNAandRNA_withanchor.py SRR13248069.DNA.bed SRR13248069.RNA.yes.bed SRR13248069.DNARNA.bedpe

11. Remove the PCR DNA–RNA pairs.

    $ awk -v flt=20 '$8>=flt' SRR13248068.DNARNA.bedpe | sort --parallel=10 -k14,14 -k13,13 -k1,1 -k2,2n -k3,3n -k4,4 -k5,5n -k6,6n -k8,8n -k9,9 -k10,10 -k11,11 -k12,12 > SRR13248068.DNARNA.sorted.bedpe

    $ python uniqread.py SRR13248068.DNARNA.sorted.bedpe > SRR13248068.DNARNA.uniq.bedpe

    $ awk -v flt=20 '$8>=flt' SRR13248069.DNARNA.bedpe | sort --parallel=10 -k14,14 -k13,13 -k1,1 -k2,2n -k3,3n -k4,4 -k5,5n -k6,6n -k8,8n -k9,9 -k10,10 -k11,11 -k12,12 > SRR13248069.DNARNA.sorted.bedpe

    $ python uniqread.py SRR13248069.DNARNA.sorted.bedpe > SRR13248069.DNARNA.uniq.bedpe

12. Merge the DNA–RNA pair into DNA–RNA interaction if these pairs have the same peak and RNA mark. Count the number of peak and RNA.

    $ cat SRR13248068.DNARNA.uniq.bedpe SRR13248069.DNARNA.uniq.bedpe > ChRD.DNARNA.uniq.bedpe

    $ python PetClusterWithGivenAnchors.py MH63.H3K4me3.exd.peak genome.info ChRD.DNARNA.uniq.bedpe | sort --parallel=10 -k1,1 -k4,4 -k2,2n -k5,5n -k3,3n -k6,6n > ChRD.DNARNA.givenanchor.cluster

    $ awk '$9>1' ChRD.DNARNA.givenanchor.cluster > ChRD.DNARNA.givenanchor.gt1.cluster

13. Calculate the significance of each DNA–RNA interaction by hypergeometric distribution test. All high-confidence DNA–RNA interactions were obtained.

    $ tmp='wc -l ChRD.DNARNA.uniq.bedpe | awk '{print $1}''

    $ Rscript hypergeometric5.r ChRD.DNARNA.givenanchor.gt1.cluster ChRD.DNARNA.givenanchor.gt1.cluster.withpvalue

    $ awk '$13+0.0<=0.05' ChRD.DNARNA.givenanchor.gt1.cluster.withpvalue > ChRD.DNARNA.givenanchor.FDRfiltered.txt

4 Notes

1. The bridge linker oligos are synthesized from Sangon Biotech and purified through HPLC. The phosphorylated 5′ end of forward strand oligos are adenylated and then annealed with the reverse strand oligos.
2. The proteins in the iBlock buffer are easily deteriorated. Please make sure it is normal before use and avoid any pollution into the buffer.
3. Cut the sample into small pieces on ice in order to avoid any degradation.
4. Conduct the step of adding formaldehyde in the fume hood. During the vacuuming process, the sample might move up to float. And in order to cross-link the sample successfully, the sample needs to be immersed in solution.
5. The sonication parameters should be optimized before this step.
6. The 30 uL input will be de-crosslinked by adding 5 μL Tris-HCl (1 M, pH 7.5), 2 μL EDTA (0.5 M), 10 μL SDS (10%) and 253 μL water, mix the solution well. Then add 5 μL of proteinase K (10 mg/mL) and 6 μL of 5 M NaCl to the sample, mix the solution well and incubate in 65 °C overnight, followed by DNA extraction and chromatin length test.
7. AMPure® XP beads are stored in 4 °C. Please remain the beads at room temperature for 30 min before use. Take 1.4 volume of the beads into the sample and mix well, incubate at room temperature for 5 min. Then keep the sample on a magnetic frame for 2 min until the supernantant is clear and wash the beads with 80% ethanol twice without disturbing the beads. Air-dry the beads and elute the digested DNA.
8. In order to make higher utilization of sequences, the number of cycles in PCR should not exceed 12.

9. Do not discard the supernatant.
10. Keep the tube on a magnetic frame and do not disturb the beads.
11. The beads should not be over-dry.
12. The ChRD-PET library length is short. Many ChRD-PET reads have the Tn5 adaptor sequence. These adaptor sequence need to be removed.
13. The upstream of the bivalent linker is connected with RNA, and the downstream is connected with DNA.
14. Some reads which are unmapped by HISAT2 can be aligned to the genome by Bowtie2.

References

1. Di Stefano M, Cavalli G (2022) Integrative studies of 3D genome organization and chromatin structure. Curr Opin Struct Biol 77: 102493–102501
2. Li X, Fu XD (2019) Chromatin-associated RNAs as facilitators of functional genomic interactions. Nat Rev Genet 20:503–519
3. Deng X, Song XW, Wei LY et al (2016) Epigenetic regulation and epigenomic landscape in rice. Natl Sci Rev 3:309–327
4. Ernst J, Kellis M (2010) Discovery and characterization of chromatin states for systematic annotation of the human genome. Nat Biotechnol 28:817–825
5. Sridhar B, Rivas-Astroza M, Nguyen TC et al (2017) Systematic mapping of RNA-chromatin interactions in vivo. Curr Biol 27:602–609
6. Yan ZM, Huang NM, Wu WX et al (2019) Genome-wide colocalization of RNA-DNA interactions and fusion RNA pairs. Proc Natl Acad Sci USA 116:3328–3337
7. Li X, Zhou B, Chen L et al (2017) GRID-seq reveals the global RNA-chromatin interactome. Nat Biotechnol 35:940–950
8. Bell JC, Jukam D, Teran NA et al (2018) Chromatin-associated RNA sequencing (ChAR-seq) maps genome-wide RNA-to-DNA contacts. elife 7:27024–27051
9. Bonetti A, Castelo-Branco G (2019) RADICL-seq identifies cell-type specific RNA-chromatin interactions in mouse oligodendrocyte progenitor cells. Nat Commun 67:293–294
10. Gavrilov AA, Zharikova AA, Galitsyna AA et al (2020) Studying RNA-DNA interactome by Red-C identifies noncoding RNAs associated with various chromatin types and reveals transcription dynamics. Nucleic Acids Res 48: 6699–6714
11. Gavrilov AA, Sultanov RI, Magnitov MD et al (2022) RedChIP identifies noncoding RNAs associated with genomic sites occupied by Polycomb and CTCF proteins. Proc Natl Acad Sci USA 119:2116222119–2116222121
12. Quinodoz SA, Jachowicz JW, Bhat P et al (2021) RNA promotes the formation of spatial compartments in the nucleus. Cell 184:5775–5790
13. Zhao L, Xie L, Zhang Q et al (2020) Integrative analysis of reference epigenomes in 20 rice varieties. Nat Commun 11:2658–2673
14. Peng Y, Xiong D, Zhao L et al (2019) Chromatin interaction maps reveal genetic regulation for quantitative traits in maize. Nat Commun 10:2632–2642
15. Zhao L, Wang SQ, Cao ZL et al (2019) Chromatin loops associated with active genes and heterochromatin shape rice genome architecture for transcriptional regulation. Nat Commun 10(10):3640–3652
16. Zhang Q, Guan P, Zhao L et al (2020) Asymmetric epigenome maps of subgenomes reveal imbalanced transcription and distinct evolutionary trends in Brassica napus. Mol Plant 14:604–619
17. Deng L, Gao BB, Zhao L et al (2022) Diurnal RNAPII-tethered chromatin interactions are associated with rhythmic gene expression in rice. Genome Biol 23:7–28
18. Xiao Q, Huang XY, Zhang Y et al (2022) The landscape of promoter-centred RNA-DNA interactions in rice. Nat Plants 8:157–170
19. Anders S, Pyl PT, Huber W (2015) HTSeq–a Python framework to work with

high-throughput sequencing data. Bioinformatics 31:166–169
20. Ramirez F, Dundar F, Diehl S, Gruning BA, Manke T (2014) deepTools: a flexible platform for exploring deep-sequencing data. Nucleic Acids Res 42:W187–W191
21. Martin M (2011) Cutadapt removes adapter sequences from high-throughput sequencing reads. EMBnetjournal 17:10–12
22. Li H, Handsaker B, Wysoker A, Fennell T, Ruan J, Homer N, Marth G, Abecasis G, Durbin R, Genome Project Data Processing S (2009) The sequence alignment/map format and SAMtools. Bioinformatics 25:2078–2079
23. Langmead B, Salzberg SL (2012) Fast gapped-read alignment with Bowtie 2. Nat Methods 9: 357–359
24. Kim D, Langmead B, Salzberg SL (2015) HISAT: a fast spliced aligner with low memory requirements. Nat Methods 12:357–360
25. Li H, Durbin R (2009) Fast and accurate short read alignment with Burrows-Wheeler transform. Bioinformatics 25:1754–1760
26. Magoc T, Salzberg SL (2011) FLASH: fast length adjustment of short reads to improve genome assemblies. Bioinformatics 27:2957–2963
27. Bolger AM, Lohse M, Usadel B (2014) Trimmomatic: a flexible trimmer for Illumina sequence data. Bioinformatics 30:2114–2120
28. Quinlan AR, Hall IM (2010) BEDTools: a flexible suite of utilities for comparing genomic features. Bioinformatics 26:841–842
29. Liu T (2014) Use model-based analysis of ChIP-Seq (MACS) to analyze short reads generated by sequencing protein-DNA interactions in embryonic stem cells. Methods Mol Biol 1150:81–95
30. Pertea G, Pertea M (2020) GFF Utilities: GffRead and GffCompare. F1000Res 9:304

Chapter 5

Prediction of RNA Joint Secondary Structures Based on Integer Programming

Yuki Kato and Kengo Sato

Abstract

Predicting a joint secondary structure formed by two RNAs plays a key role in understanding mechanisms and functions of complex RNA–RNA interactions. Here we describe a practical procedure of a fast and accurate computational method for predicting joint structures based on integer programming.

Key words RNA–RNA interaction, RNA joint secondary structure, Regulatory RNA, Integer programming, Maximum expected accuracy

1 Introduction

A noncoding RNA can interact with its target RNA by base pairing that mainly consists of Watson–Crick {A, U} and {G, C} pairs along with wobble {G, U} pairs. This string-based rule has motivated computational scientists to develop a variety of algorithms for predicting RNA–RNA interactions. They are classified into two approaches: one that neglects intramolecular base pairs and the other that considers internal structures of two RNAs [1]. Most of the latter approaches aim to predict a joint secondary structure, which tends to reflect more realistic interactions than the former approach that mainly computes thermodynamic scores of duplexes between two RNAs. In general, predicting a joint structure with minimum free energy is NP-hard if there is no constraint on the topology [2]. Thus, adding valid constraints on the topology of predictable joint structures is reasonable, and existing methods impose such constraints. For example, intramolecular crossing base pairs (pseudoknot) and external crossing interactions are disallowed. An example of complicated but restricted class of RNA–RNA interactions is a kissing hairpin, where loops of hairpin substructures from two RNAs are interacted with each other (Fig. 1). Some computational methods can predict kissing hairpins [2–6],

Michiaki Hamada et al. (eds.), *RNA-RNA and RNA-DNA Interactions*, Methods in Molecular Biology, vol. 2949,
https://doi.org/10.1007/978-1-0716-4670-0_5,

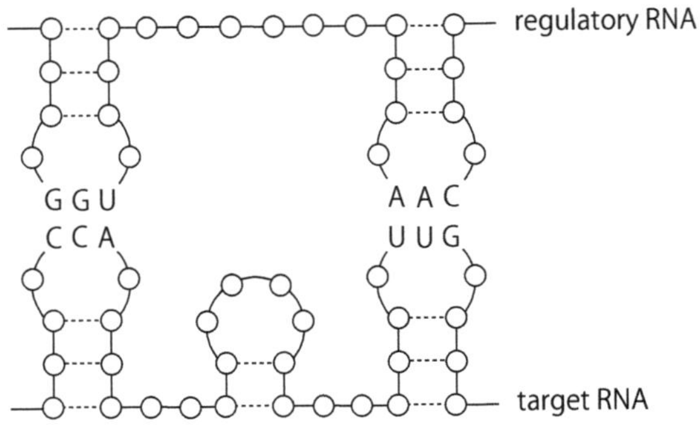

Fig. 1 An example of two kissing hairpin structures formed by a regulatory RNA and its target. A white circle indicates a base, and a dashed line shows an internal base pair. To emphasize external base pairs underling the RNA–RNA interaction, actual bases in the hairpin loop regions are shown to illustrate Watson–Crick base pairings

but their computation time is $O(n^6)$, where n is the sequence length.

To overcome such high computational complexity, RactIP (RNA–RNA interaction prediction using integer programming) was proposed on the basis of an integer programming (IP) formulation [7]. In an IP problem, we would like to minimize/maximize a linear objective function subject to linear constraints described by linear equalities and/or inequalities with respect to variables of nonnegative integers. Notably, modeling with IP is flexible enough to solve a wide range of combinatorial problems, and recent optimization solvers can solve IP problems faster than previously thought. In the RactIP formulation, an approximate posterior probability of a joint secondary structure is incorporated into the objective function of its IP formulation. Maximizing this objective function corresponds to maximizing expected accuracy of a predicted joint secondary structure. Such approximation achieves considerable speedup in computation time while keeping prediction accuracy [7]. The topology of predictable joint structures can be modeled by a set of constraints. Furthermore, motivated by some accessibility-based approaches that compute a free energy to make a sequence stretch open for RNA–RNA binding [8, 9], RactIPAce (RactIP with precomputed accessibility) was developed to incorporate interaction site accessibility to improve prediction accuracy [10].

In what follows, we will describe a practical procedure to install and run RactIP (RactIPAce) for predicting joint structures based on IP. Note that RactIP includes RactIPAce because the function of RactIPAce can be invoked as a command line option when using the latest version of RactIP.

2 Materials

2.1 Software

RactIP can run by command line interface on Linux, Windows Subsystem for Linux (WSL), and macOS. Before installing RactIP onto user's computer, a few tools described below need to be installed. Although we illustrate detailed procedures on Linux (Ubuntu 24.04.3 LTS) as an example, we believe that these can be performed similarly in other computational environments. In the following, assume that we are in `/home/username` directory.

2.1.1 ViennaRNA Package

The ViennaRNA package [11] version 2.7.0 or earlier, a toolkit for analyzing RNA secondary structures, is required by RactIP. The ViennaRNA package can be installed in your terminal by the following steps:

Command

```
wget https://www.tbi.univie.ac.at/RNA/download/\
sourcecode/2_7_x/ViennaRNA-2.7.0.tar.gz
tar zxf ViennaRNA-2.7.0.tar.gz
cd ViennaRNA-2.7.0
./configure
make
sudo make install
cd ..
```

The above commands will install the ViennaRNA package to `/usr/local` (*see* **Note 1**).

2.1.2 Optimization Solvers

Any of the optimization solvers, GNU Linear Programming Kit (GLPK) ($\geq$ 4.41) [12], Gurobi Optimizer ($\geq$ 8.0) [13], IBM ILOG CPLEX ($\geq$ 12.0) [14], SCIP ($\geq$ 8.0.3) [15], and HiGHS ($\geq$ 1.5.0) [16], is necessary to install to solve IP problems designed in the RactIP framework. For example, the free package GLPK 5.0 can be installed by the commands shown below (*see* **Note 2**):

Command

```
wget https://ftp.gnu.org/gnu/glpk/glpk-5.0.tar.
gz
tar zxf glpk-5.0.tar.gz
```

(continued)

```
cd glpk-5.0
./configure
make
sudo make install
cd ..
```

Refer to an instruction of another solver to install it (*see* **Note 3**).

2.2 Installing RactIP

2.2.1 Building an Executable Program from Source Codes

We demonstrate how RactIP 1.1.1 can be built with GLPK. Make sure that a C++17 compatible compiler and the commands `cmake` and `pkg-config` work on your environment; otherwise install them. The GitHub website at https://github.com/satoken/ractip is also helpful to install RactIP.

Command

```
wget https://github.com/satoken/ractip/archive/\
refs/tags/v1.1.1.tar.gz
tar zxf v1.1.1.tar.gz
cd ractip-1.1.1
export PKG_CONFIG_PATH=/usr/local/lib/pkgcon-
fig:\
$PKG_CONFIG_PATH
cmake -DCMAKE_BUILD_TYPE=Release -S . -B build
cmake --build build
sudo cmake --install build
```

For Gurobi, add `-DENABLE_GUROBI` to the above configure step:

Command

```
cmake -DENABLE_GUROBI=true \
      -DCMAKE_BUILD_TYPE=Release -S . -B build
```

For CPLEX, add `-DENABLE_CPLEX` to the configure step (*see* **Note 4**):

Command

```
cmake -DENABLE_CPLEX=true \
      -DCMAKE_BUILD_TYPE=Release -S . -B build
```

For SCIP, add -DENABLE_SCIP to the configure step (*see* **Note 5**):

Command

```
cmake -DENABLE_SCIP=true \
      -DCMAKE_BUILD_TYPE=Release -S . -B build
```

For HiGHS, add -DENABLE_HIGHS to the configure step:

Command

```
cmake -DENABLE_HIGHS=true \
      -DCMAKE_BUILD_TYPE=Release -S . -B build
```

2.2.2 Using a Precompiled Binary

Users can also install RactIP precompiled binary if they use Linux or WSL. Assume that we are in the /home/username directory.

Command

```
wget https://github.com/satoken/ractip/releases/\
download/v1.1.1/ractip-1.1.1-x86_64-linux.zip
unzip ractip-1.1.1-x86_64-linux.zip
```

The executable file ractip can be found in the ractip-1.1.1-x86_64-linux directory.

3 Methods

3.1 Data Preparation

Example input sequences in FASTA format for performing joint secondary structure prediction are available in `ractip-1.1.1/data` directory or `ractip-1.1.1-x86_64-linux/data`, which have already been downloaded through the installation of RactIP described above.

3.2 Predicting Joint Secondary Structures

The current directory is assumed to be `/home/username/ractip-1.1.1`. Taking sequences `CopA.fa` and `CopT.fa` in FASTA format (*see* **Note 6**), which are known to form a kissing hairpin structure [17], RactIP can run by the following command:

Command

```
ractip data/CopA.fa data/CopT.fa
```

A predicted joint secondary structure will soon be returned in dot-bracket representation to standard out as shown in Fig. 2. To evaluate the prediction quantitatively, RactIP has the option `-e` to calculate the free energy of the predicted joint structure. For example, free energy − 37.88 kcal/mol of the joint structure formed by CopA–CopT will be shown at the bottom of the prediction.

3.3 Running with Docker

If Docker has been installed, the following commands can run RactIP:

Command

```
docker build . -t ractip
docker run -it --rm -v $(pwd):$(pwd) -w $(pwd) \
ractip ractip CopA.fa CopT.fa
```

```
>CopA
CGGUUUAAGUGGGCCCCGGUAAUCUUUUCGUACUCGCCAAAGUUGAAGAAGAUUAUCGGGGUUUUUGCUU
.......(((.((((((((((((((((.[[[[[[[[[[[[[[[....))))))))))))))))...))).
>CopT
AAGCAAAAACCCCGAUAAUCUUCUUCAACUUUGGCGAGUACGAAAAGAUUACCGGGGCCCACUUAAACCG
..........(((((.(((((((.....]]]]]]]]]]]]]]]..))))))).))))).............
```

Fig. 2 A predicted kissing hairpin structure formed by antisense RNA CopA and its target RNA CopT. Parentheses "( )" denote an internal base pair, whereas brackets "[ ]" show an external base pair

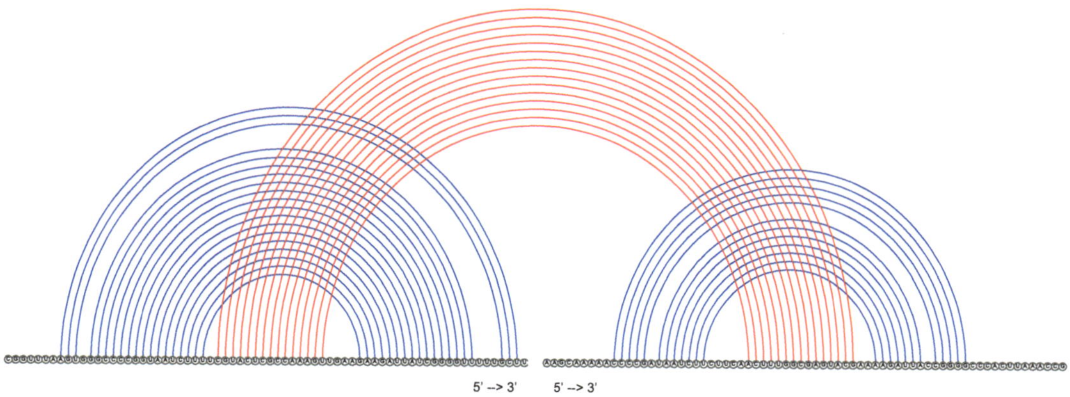

Fig. 3 An arc representation of the predicted joint structure formed by CopA and CopT. Note that both sequences are oriented in the 5′ to 3′ direction from left to right. A blue arc indicates an intramolecular base pair, and a red arc shows an interaction between two RNAs

3.4 Using a Web Server

To avoid installation for some reason or see an arc representation of a predicted joint structure explained below, we provide a web server Rtips [18] online at http://ws.sato-lab.org/rtips/ractip/. In the web server, CopA and CopT sequences will appear as Example 3 in the top page, and just clicking "Predict" button will perform joint structure prediction as shown in the command line case. Unlike the standalone program, the web server will output not only a dot-bracketed joint structure with the predicted free energy but also an arc representation of the prediction shown in Fig. 3 by virtue of a drawing tool VARNA [19].

4 Notes

1. A user without root privileges on his/her computer cannot edit `/usr/local`, which means that the final command `sudo make install` will fail. To resolve this issue, create a directory under the home directory such as `/home/username/ViennaRNA`, and run the configure script as follows:

```
./configure --prefix=/home/username/ViennaRNA
```

2. A non-root user should install GLPK by the command like this:

```
./configure --prefix=/home/username/glpk
```

where the directory `glpk` has been made under the home directory.

3. Gurobi Optimizer and IBM ILOG CPLEX are commercial software but grant free academic license.
4. If the optimization solver has been installed to the directory other than `/usr` or `/usr/local`, add an option that indicates the root directory of the solver installed to the configuration step. Here is an example of CPLEX-incorporated installation:

```
cmake -DCPLEX_ROOT_DIR=/path/to/CPLEX \
  -DENABLE_CPLEX=true -DCMAKE_BUILD_TYPE=Release \
  -S . -B build
```

   As for the case of Gurobi, add `-DGROBI_ROOT_DIR=/path/to/Gurobi` to the configuration.
5. If SCIP has been installed to the directory other than `/usr` or `/usr/local`, specify the option `-DCMAKE_MODULE_PATH=/path/to/scip/lib/cmake` in the configuration step.
6. The FASTA formatted file begins with ">" including its sequence ID, followed by an actual sequence from the second row.

Acknowledgements

This work was supported by the Japan Society for the Promotion of Science KAKENHI [24700296 and 15K00401 to YK and 16K00404 to KS].

References

1. Backofen R (2014) Computational prediction of RNA–RNA interactions. Methods Mol Biol 1097:417–435
2. Alkan C, Karakoç E, Nadeau JH et al (2006) RNA–RNA interaction prediction and antisense RNA target search. J Comput Biol 13: 267–282
3. Pervouchine DD (2004) IRIS intermolecular RNA interaction search. Genome Inform 15: 92–101
4. Chitsaz H, Salari R, Sahinalp SC, othres (2009) A partition function algorithm for interacting nucleic acid strands. Bioinformatics 25:i365–i73
5. Huang FWD, Qin J, Reidys CM, othres (2009) Partition function and base pairing probabilities for RNA–RNA interaction prediction. Bioinformatics 25:2646–2654
6. Kato Y, Akutsu T, Seki H (2009) A grammatical approach to RNA–RNA interaction prediction. Pattern Recognit 42:531–538
7. Kato Y, Sato K, Hamada M et al (2010) RactIP: fast and accurate prediction of RNA–RNA interaction using integer programming. Bioinformatics 26:i460–i466
8. Mückstein U, Tafer H, Hackermüller J et al (2006) Thermodynamics of RNA–RNA binding. Bioinformatics 22:1177–1182
9. Busch A, Richter AS, Backofen R (2008) IntaRNA: efficient prediction of bacterial sRNA targets incorporating target site accessibility and seed regions. Bioinformatics 24: 2849–2856
10. Kato Y, Mori T, Sato K et al (2017) An accessibility-incorporated method for accurate prediction of RNA–RNA interactions from sequence data. Bioinformatics 33:202–209
11. Lorenz R, Bernhart SH, Höner Zu Siederdissen C, et al (2011) ViennaRNA package 2.0. Algorithms Mol Biol 6:26
12. GLPK (GNU Linear Programming Kit). https://www.gnu.org/software/glpk/

13. Gurobi Optimization. https://www.gurobi.com/
14. IBM ILOG CPLEX Optimization Studio. https://www.ibm.com/products/ilog-cplex-optimization-studio
15. SCIP (Solving Constraint Integer Programs). https://scipopt.org/
16. HiGHS (high performance software for linear optimization). https://highs.dev/
17. Kolb FA, Malmgren C, Westhof E et al (2000) An unusual structure formed by antisense-target RNA binding involves an extended kissing complex with a four-way junction and a side-by-side helical alignment. RNA 6:311–324
18. Kato Y, Sato K, Asai K et al (2012) Rtips: fast and accurate tools for RNA 2D structure prediction using integer programming. Nucleic Acids Res 40:W29–W34
19. Darty K, Denise A, Ponty Y (2009) VARNA: interactive drawing and editing of the RNA secondary structure. Bioinformatics 25:1974–1975

Chapter 6

ADAR1 RNA-Editing Enzyme Regulates Telomeric R-Loop Formation

Eito Ichihashi, Mai Kubota, Yusuke Shiromoto, and Masayuki Sakurai

Abstract

The enzyme adenosine deaminase acting on RNA (ADAR) plays a key role in editing RNA by converting adenosine to inosine, a process known as A-to-I RNA editing. Initially, ADAR was identified as an enzyme responsible for converting adenosine to inosines, specifically in dsRNA. However, recent studies indicate that ADAR1 can also edit DNA on hybrid strands composed of DNA:RNA strands. One of the representative structures in cells is the R-loop, which is formed at open DNA regions with single-stranded DNA and DNA:RNA hybrid strands, through hybridization of the cis-transcript or an invading RNA strand with the template DNA. In the previous study, authors have shown that ADAR1 regulates the R-loop formation and genome stability at telomeres in cancer cells carrying noncanonical variants of telomeric repeats.

In this chapter, the latest and basic protocols to measure the accumulation and formation of the R-loop are described.

Key words DNA:RNA, R-loop, S9.6 antibody, Dot blot, Antibody, Genomic DNA, ADAR

1 Introduction

In general, the genomic DNA comprises a double-stranded form. An R-loop structure has been found as one of the noncanonical structures on the genomic DNA, which is formed at opened DNA regions with single-stranded DNA and DNA:RNA hybrid strands by staying of the cis-transcript or RNA trans-intrusion. The R-loop formation has been reported universally among species, from bacteria to mammals. R-loops are implicated in various biological processes, such as gene regulation, DNA repair, class switch recombination in the immune response, and replication fork stabilization [1–4]. However, R-loops can also be sources of genomic instability. If not properly resolved, R-loops can cause conflicts between transcription and replication machinery, leading to DNA damage and potentially harmful mutations. This is why cells have specific proteins that regulate R-loop formation and resolution to maintain

Michiaki Hamada et al. (eds.), *RNA-RNA and RNA-DNA Interactions*, Methods in Molecular Biology, vol. 2949,
https://doi.org/10.1007/978-1-0716-4670-0_6,

genomic integrity. The study of R-loops is an active area of research, and there is still much to learn about their formation, resolution, and biological functions. Their roles in health and disease are also being explored, including links to neurodegenerative diseases and cancer.

R-loops tend to form preferentially at regions with high GC content in the genome. The stability of the DNA:RNA hybrid in these regions is higher than that of the DNA:DNA double helix, favoring the formation of R-loops. In regions of the genome with positive GC-skew (more guanines than cytosines), R-loops are more likely to form. GC-skew refers to the asymmetry in the distribution of guanines and cytosines on the two strands of DNA. It is calculated as $(G - C)/(G + C)$.

Certain DNA sequences and structures are more prone to R-loop formation. This includes regions with repetitive sequences and sequences that can form secondary structures (like G-quadruplexes). The dynamics of the transcription machinery can also influence R-loop formation. For instance, a slow-moving or stalled RNA polymerase might favor R-loop formation. In addition, R-loops are more likely to form during the transcription of genes that are highly expressed. Approximately 34% of R-loops are located in coding regions, while around 8% are found in promoter regions. They are also frequently formed in repetitive sequences, such as long interspersed nuclear elements (LINE), short interspersed nuclear elements (SINE), as well as tRNA and rRNA gene loci [5].

The R-loop is known to act as a transcriptional activator or repressor at specific loci. It can also promote heterochromatin formation by facilitating the compression of repetitive sequences [6–8]. R-loops accumulate at enhancers, promoters, and terminators of activated genes and typically overlap with histone markers and histone acetylation. Otherwise, it has also been associated with repressive histone marks and DNA methylation [9, 10]. However, the detailed mechanism by which R-loop acts at a particular locus remains unknown. Therefore, substantial attention focused on the study of R-loops has been growing.

Furthermore, our laboratory has shown that ADAR1, an RNA-editing enzyme, binds to and edits the R-loop, the DNA: RNA hybrid strand. We have reported that ADAR1 regulates R-loop and genome stability when telomere R-loop formation occurs in cancer cells with non-normal mutants of telomere repeats [12]. ADAR1p110 edits A–C mismatches within the DNA:RNA hybrid formed between canonical and noncanonical variant repeats. Moreover, ADAR1p110-mediated editing of A–C-mismatched base pairs, which converts them to I:C-matched base pairs, is required for the degradation of the RNA strands of telomeric repeat DNA:RNA hybrids by RNase H2. Hence, ADAR1p110-dependent regulation of the telomere R-loop is necessary for telomerase-activated cancers to continue proliferating, and it is required for the continued proliferation of telomerase-inactivated cancer cells, revealing the oncogenic potential of ADAR1p110 [12] In these studies, we used the S9.6 antibody to detect R-loops, which binds specifically to DNA:RNA hybrids and is widely used for mapping R-loops [12]. The S9.6 Dot Blot was used to analyze the amount of R-loop in the whole genome of ADAR1 knockdown cells. Furthermore, we performed DRIP (DNA:RNA-Immunoprecipitation) using an S9.6 antibody, followed by qPCR to quantify the amount of R-loops in specific genes. In addition, using each Dot-Blot hybridization probe on the DRIP samples makes it possible to compare the R-loop amounts for a specific gene range [12]. Thus, detection using dot blot and DRIP with S9.6 allows visualization and quantification of changes in the amount of DNA:RNA hybrid strands in each cell, thereby enabling the study of biological phenomenon related to R-loop structures.

In this chapter, we describe several experiments to analyze the amount of R-loops using the S9.6 antibody, which has been reported in Shiromoto et al. [12]. We investigated the relationship between ADAR1 and R-loops in the telomere region by following the experimental steps outlined in Fig. 1, from steps 1–6. In this chapter, we describe the first two steps of those experiments.

2 Materials

2.1 DNA Recovery from Cells

1. Triple Express reagent (Thermo Fisher Science).
2. Blood and Cell Culture DNA Midi Kit (QIAGEN).
3. PBS: 200 mg/L of potassium chloride, 200 mg/L of potassium dihydrogen phosphate, 8000 mg/L of sodium chloride, and 1150 mg/L of disodium hydrogen phosphate.
4. 2-propanol.
5. 70% ethanol.
6. Centrifuge with a swinging rotor.

2.2 Dot Blot

1. Anti-DNA:RNA hybrid antibody [S9.6] (Abcam).
2. Peroxidase AffiniPure donkey anti-mouse IgG (H + L) (Jackson ImmunoResearch) (*see* **Note 1**).
3. Chemi-Lumi One L (Nacalai Tesque).
4. Methylene Blue.
5. Sense 62 nt RNA (Table 1): (5′→3′) GGCGAGAGGCAA GAGGCGCGCAGUAGGGCGGCAGAAGCGGC GUAGCGGGCCGCGCGUCGGGC.

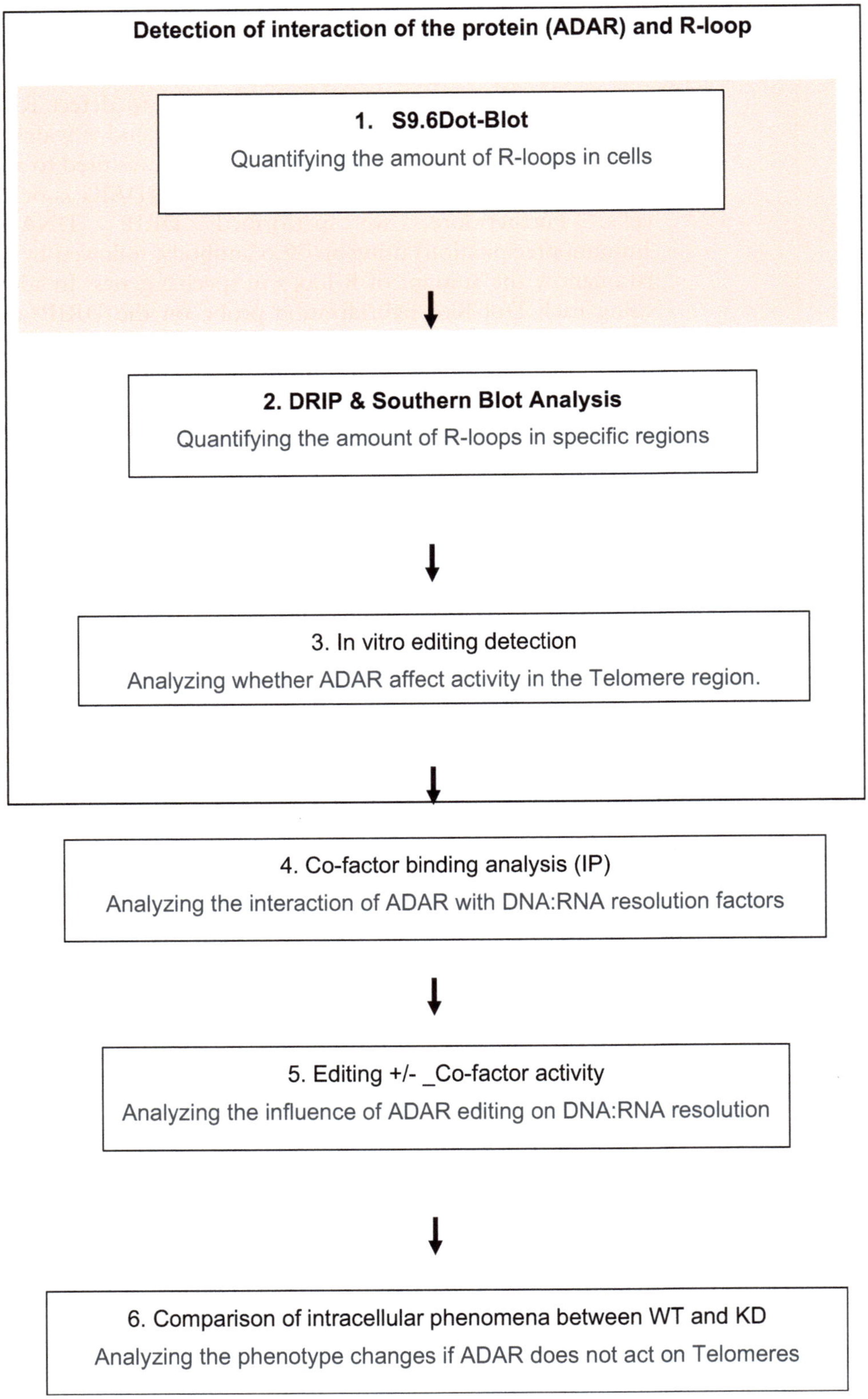

Fig. 1 Steps involved in the detection process. This figure depicts the necessary steps (1–6) required to investigate the relationship between ADAR1 and R-loops in the telomere region [12]. This chapter primarily describes the first two steps of these experiments

Table 1
Combinations of sense oligo and antisense oligo for the controls

Control sets for S9.6	Sense Oligo	Antisense Oligo
DNA:RNA positive control	Sense 62 nt RNA	Antisense 62 nt DNA
DNA:DNA negative control	Sense 62 nt DNA	Antisense 62 nt DNA
RNA:RNA negative control	Sense 62 nt RNA	Antisense 62 nt RNA

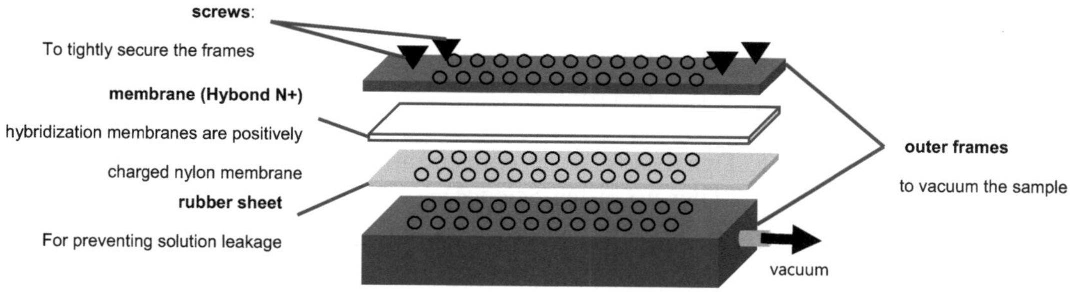

Fig. 2 Bio-Dot apparatus

6. Antisense 62 nt RNA: (5′→3′) GCCCGACGCGCGGCCCG CUACGCCGCUUCUGCCGCCCUACUGCGCGCCU CUUGCCUCUCGCC.
7. Sense 62 nt DNA: (5′→3′). GGCGAGAGGCAAGAGGCGCGCAGTAGGGCGG-CAGAAGCGGCGTAGCGGGCCGCGCGTCGGGC.
8. Antisense 62 nt DNA: (5′→3′). GCCCGACGCGCGGCCCGC-TACGCCGCTTCTGCCGCCC-TACTGCGCGCCTCTTGCCTCTCGCC.
9. Annealing buffer: 10 mM Tris–HCl (pH 7.4), 50 mM NaCl.
10. 6× SSC: Dilute the SSC buffer stock solution (20×) (Nacalai Tesque) with water to achieve the required concentration.
11. PBST (0.05%): To prepare 200 mL of PBST, add 1 mL of 10% Tween 20 to phosphate-buffered saline (PBS).
12. 2× BSA/PBST (0.2%): To prepare 2× BSA/PBST, add 10 mL of Blocker BSA (10×) in PBS and 1 mL of 10% Tween 20–40 mL of PBS.
13. 5% Skim milk/PBST (0.1%): 5% skim milk/PBST (0.1%): To prepare 50 mL of 5% skim milk/PBST, add 500 μL of 10% Tween 20 and 2.66 g of skim milk (Nacalai Tesque) to PBS. Dissolve the skim milk by inverting the tube (*see* **Note 2**).

14. 1× BSA/PBST (0.05%): To prepare 1× BSA/PBST, add 2 mL of Blocker BSA (10×) in PBS and 100 μL of 10% Tween 20 to 18 mL of PBS.
15. 0.3 M sodium acetate (pH 5.2)
16. Hybond N+ membrane (Cytiva).
17. Bio-Dot Apparatus (Biorad).
18. UV cross-linker (UVP).
19. For shaking needs, use a reciprocating shaker (TAITEC).
20. iBright Imaging System (Thermo Fisher Science).
21. Nanodrop One (Thermo Fisher Science).
22. Aspirator (BIOSAN).

2.3 Drip

1. Dynabeads at a concentration of 30 mg/mL Protein A (Thermo Fisher Science).
2. Normal Mouse IgG Antibody (Santa Cruz).
3. RNA/DNA grade Proteinase K (20 mg/mL) (Thermo Fisher Science).
4. RNaseInPlus Inhibitor (Promega).
5. RNase H (New England Biolabs).
6. DynaMag™-2 (Thermo Fisher Science).
7. MinElute PCR Purification Kit (QIAGEN).
8. D buffer: 10 mM Tris–HCl (pH 8.5), 300 mM NaCl.
9. Blocking solution consisting of 1× PBS-5 mM EDTA-0.5% BSA.
10. 2× DRIP buffer: 100 mM Tris–HCl (pH 7.4), 10 mM EDTA, 2.0% NP-40, 0.2% sodium deoxycholate.
11. 2× DNA dilution DRIP buffer: 100 mM Tris–HCl (pH 7.4), 300 mM NaCl, 10 mM, EDTA, 2.0% NP-40, 0.2% sodium deoxycholate.
12. 1× DRIP buffer (RNaseIn): 50 mM Tris—HCl (pH 7.4), 150 mM NaCl, 5 mM EDTA, 1.0% NP- 40, 0.1% sodium deoxycholate, 40 U/μL RNaseInPlus (Promega).
13. DRIP H buffer: 50 mM Tris–HCl (pH 7.4), 500 mM NaCl, 5 mM EDTA, 1.0% NP-40, 0.1% sodium deoxycholate.
14. DRIP Li buffer: 50 mM Tris–HCl (pH 8.0), 250 mM LiCl, 5 mM EDTA, 1.0% NP-40, 0.1% sodium deoxycholate.
15. TE + Salt (50 mM NaCl) buffer: 100 mM Tris–HCl (pH 8.0), 10 mM EDTA, 50 mM NaCl.
16. TE buffer: 100 mM Tris–HCl (pH 8.0), 10 mM EDTA.
17. Elution buffer: 50 mM Tris–HCl, pH 8.0, 10 mM EDTA, 1.0% SDS.

18. RNase H buffer: 25 mM Tris–HCl (pH 7.5), 30 mM NaCl, 0.5 mM EDTA, 1 mM DTT, 50% glycerol.
19. 0.13 N NaOH/3.3 mM EDTA.
20. 2× SSC with 0.1% SDS solution.
21. ULTRAhyb Ultrasensitive Hybridization Buffer (Thermo Fisher Science).
22. 5′-^{32}P-labeled DNA oligonucleotide probe.
23. Sonicator (VibraCell).
24. Typhoon RGB Imager (GE Healthcare).
25. Eppendorf ThermoMixer® C(Eppendorf).

3 Methods

3.1 DNA Recovery from Cells

Avoid adding RNaseA to BufferG2, as it will prevent the recovery of DNA:RNA hybrid strands.

1. Remove the supernatant from the T75 flask and add 3 mL of Triple Express (use a volume of Triple Express that is one-fifth of the total medium volume).
2. Incubate at 37 °C for 5 min, then add 12 mL of cold PBS, pre-chilled to 4 °C, and mix well.
3. Transfer the total 15 mL of the suspended solution to a 50 mL Falcon tube and centrifuge in a swinging rotor at 700× *g* for 5 min at 4 °C.
4. Discard the supernatant, resuspend in 10 mL of cold PBS, and repeat the centrifugation step.
5. Again, discard the supernatant, and adjust the cell density to 1.0×10^7 cells/mL with cold PBS.
6. Add 2 mL of cold C1 and 6 mL of cold water, then gently vortex to create a suspension (*see* **Note 3**).
7. Centrifuge the suspension at 1300× *g* for 20 min at 4 °C using a swinging rotor.
8. Discard the supernatant, then add 2 mL of cold C1 and 6 mL of cold water, and lightly vortex to create a suspension.
9. Centrifuge the suspension at 1300× *g* for 15 mins at 4 °C, then discard the supernatant.
10. Add 5 mL of BufferG2 and gently vortex to suspend the pellet.
11. Add 95 μL of QIAGEN Proteinase and incubate at 50 °C for 30–60 min.
12. Add 4 mL of BufferQBT to the genomic tip and allow it to pass through via gravity flow.

13. Add 7.5 mL of BufferQC to the genomic tip and let it pass through via gravity flow; repeat this step.
14. Add 5 mL of BufferQF to the genomic tip to elute DNA into a 50-mL Falcon tube.
15. Add 3.5 mL of room temperature 2-propanol and vortex to create a suspension.
16. Centrifuge this tube at 15,000× *g* for 15 min at 4 °C using a swinging rotor.
17. Carefully remove the supernatant using an aspirator, ensuring to avoid aspirating the DNA pellets, which become visible after centrifugation (*see* **Note 4**).
18. Resuspend the DNA pellets in 2 mL of cold 70% ethanol (4 ° C), transfer the suspension to a 2-mL tube, and centrifuge at 15,000× *g* for 10 min at 4 °C.
19. Discard the supernatant and add 250 μL of water.
20. Dissolve the DNA overnight on a shaker or at 55 °C for 1–2 h.

3.2 Dot Blot

3.2.1 Sample Preparation

1. Measure the concentration of the recovered DNA using the Nanodrop One.
2. Dilute the DNA in 6× SSC to achieve a concentration of 1 μg/ 100 μL. It is recommended to dilute the DNA to 5 μg/500 μL, which can conduct dot blot experiments with replicates; N = 4 per sample (*see* **Note 5**).

3.2.2 Preparation of S9.6 Dot-Blot Control Sample

1. In a 1.5-mL tube, combine 10 μL of annealing buffer, 20 pmol of Sense Oligo, 24 pmol of Antisense Oligo, and enough water to reach a total volume of 100 μL.
2. Incubate the solution at 80 °C for 5 min.
3. Gradually cool the solution to room temperature.

3.2.3 Membrane Preparation

1. Cut a positively charged nylon membrane (Hybond N+) to fit the dot blot device; in our setup, this requires dimensions of 12.5 cm length and 9 cm width.
2. Immerse the cut membrane in water and shake gently for 5 min.
3. Remove the membrane from the water, soak it in 6× SSC, and shake for another 5 min.

3.2.4 Dot-Blot Apparatus

1. Remove the outer frame by loosening the screws. Place the membrane on the rubber sheet, positioning it between the membrane and the outer frame (*see* **Note 6**).
2. Reattach the gray outer frame and secure it by tightening the screws (*see* **Note 7**).

3. Add the prepared DNA sample and 100 μL of each control sample (DNA:RNA, DNA:DNA, RNA:RNA) to the well (*see* **Note 8**).
4. Vacuum with an aspirator, allowing the solution to drop and adsorb onto the membrane (*see* **Note 9**).
5. Add approximately 300 μL of 6× SSC to the well and rinse by vacuuming with an aspirator in the same manner.

3.2.5 UV Cross-Linking

1. Allow the membrane to dry at room temperature for approximately 5–10 min, then cover it with aluminum foil to protect it from contamination and light exposure (*see* **Note 10**).
2. Irradiate the membrane with UV light at 120 mJ/cm^2 for cross-linking, ensuring that the membrane remains on the aluminum foil while in the cross-linker device (*see* **Note 11**).
3. Flip the membrane upside down and irradiate it again with UV light at 120 mJ/cm^2 to cross-link the opposite side (*see* **Note 12**).

3.2.6 Blocking

1. Soak the UV-cross-linked membrane in 5% skim milk/PBST (0.1%) and shake for 30 min (*see* **Notes 13** and **14**).
2. Remove the membrane and gently wash it once with PBST (0.05%).
3. Immerse in 20% BSA/PBST (0.2%) and shake for 30 min (*see* **Notes 13** and **14**).

3.2.7 Primary Antibody Reaction

1. Dilute the anti-DNA:RNA hybrid antibody [S9.6] to 1/4000 with 1× BSA/PBST (0.05%). Prepare 10–20 mL of the diluted antibody for each membrane.
2. Immerse the membrane in the primary antibody solution and shake it at 4 °C overnight (*see* **Notes 13** and **14**).

3.2.8 Wash and Secondary Antibody Reaction

1. Collect the membrane that has undergone the secondary antibody reaction.
2. Immerse the membrane in PBST (0.05%) and shake for 10 min to wash (*see* **Notes 13** and **14**).
3. Repeat **step 2** three more times (for a total of four times).
4. Dilute the secondary antibody to 1/10,000 with 1× BSA/PBST. Prepare 10–20 mL of diluted antibody for each membrane.
5. React the membrane with the secondary antibody at room temperature for 1 h, shaking it during this period (*see* **Notes 13** and **14**).

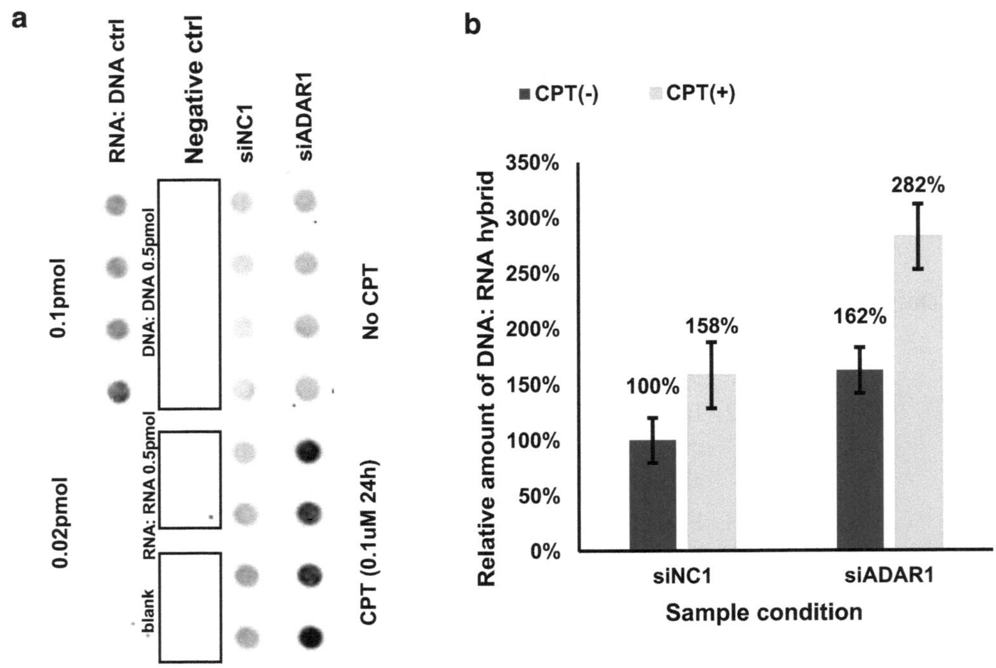

Fig. 3 Results of the S9.6 dot blot. (**a**) The S9.6 dot blot was utilized to quantify the amount of R-loops in HCT116 cells following ADAR1 knockdown. Controls included 0.1 and 0.02 pmol DNA/RNA hybrids, 0.5 pmol dsRNA, 0.5 pmol dsDNA, and blank samples. Results indicate an increase in R-loop quantities post-ADAR1 knockdown. Additionally, R-loop concentrations from DNA damage increased in samples treated with Camptothecin (CPT) (*see* **Note 21**). (**b**) This is a quantitative representation of (**a**). The black bars represent samples without CPT treatment, while the gray bars denote samples treated with CPT

3.2.9 Wash and Detection Reaction

1. Collect the membrane that has undergone the secondary antibody reaction.
2. Immerse the membrane in PBST (0.05%) and shake it for 10 min to wash.
3. Repeat **step 2** four more times (for a total of five times).
4. Wash the membrane in PBS for 10 min, and then lightly drain off the water.
5. Use a chemiluminescence detection reagent for Western blotting. In our case, we mix equal volumes of Chemi-Lumi One L Solution A and Solution B. Prepare a total of 2 mL per membrane and pour the mixture over the membrane.
6. Incubate at room temperature for 1 min.
7. Lightly drain off the solution and then proceed with detection. We use the iBright for this step (Fig. 3).

3.2.10 Confirmation of Nucleic Acid Application Content

1. To check the amount of nucleic acid applied, stain the membrane with Methylene Blue.

2. Prepare a 0.02% Methylene Blue solution in 0.3 M sodium acetate (pH 5.2).
3. Stain the membrane at room temperature for 30 min.
4. Submerge the membrane in water and agitate for 5 min to achieve destaining.
5. Repeat **step 3** two more times (for a total of three times).

3.3 Drip

3.3.1 Sonication

1. Dissolve 50 μg of DNA in 250 μL D buffer (10 mM Tris–HCl (pH 8.5), 300 mM NaCl).
2. Sonicate the DNA to fragments of 200–400 bp. In our conditions, we use VibraCell's sonicator, and use a 2-mm probe, at 35% amplitude ×10 s (with 20 s intervals) for 20 cycles. Between sonication, keep the samples on ice (*see* **Note 15**).

3.3.2 Preparing Magnetic Beads

The setting below is for 40–50 μg genomic DNA per one condition. All steps are performed at 4 °C.

In the following steps, *see* **Note 16** if it is the first time to use a magnetic beads system.

1. Add 1.5 mL of Block Solution (×1.5 volume of the starting beads) and gently resuspend the beads. You can do this by removing the tubes from the rack and inverting the tubes 10–20 times until the beads are evenly resuspended or by removing the magnetic strip from the rack and inverting the rack with the tubes still in place—either 10–20 times or until the beads are evenly resuspended.
2. Block the solution at 4 °C for 2–4 h, or overnight.
3. Collect the beads using a magnetic bead stand. Remove the supernatant with an aspirator or a pipettor.
4. Wash the beads twice in 1.5 mL of 1 × DRIP IP buffer.
5. Resuspend beads in 100–200 μL (×1–2 vol of start beads volume) in 1× DRIP IP buffer to prepare 50–25% slurry beads and add 20 μg of antibody (S9.6 or mouse IgG as negative control).
6. Incubate at 4 °C for 6 h or overnight, on a rotator.
7. Wash the beads twice in 1 mL 1× DRIP IP buffer, as described above.
8. Resuspend each aliquot of beads in 100 μL (×1 starting beads volume) in 1× DRIP IP buffer (50% slurry beads).

3.3.3 Immunoprecipitation

1. After sonication, the DNA volume (50 μg DNA/250 μL) should be reduced to 230 μL. The DNA concentration is now 50 μg/230 μL. Reserve 20 μL as a sample for input.
2. Transfer the solution to a Protein LoBind 2.0 mL Eppendorf tube (*see* **Note 17**).

3. Add 210 μL (an equal volume to the 210 μL DNA solution) of 2× DNA dilution DRIP buffer.
4. Add 100 μL of ProteinA Beads (S9.6 20 μg). The total volume should now be between 200 and 500 μL.
5. Incubate overnight on a rotator at 4 °C.

3.3.4 Wash and Elution

In the following steps, *see* **Note 18** if this is the first time to use a magnetic beads system.

For washing **steps 1–3**: Add 1 mL of the respective wash buffer to the tube. Detach the tube from the stand and mix the contents before returning it to the stand. This process should be repeated twice for each wash buffer.

1. Wash twice with 1 mL of 1× DRIP buffer (RNaseIn).
2. Wash twice with 1 mL of DRIP H buffer.
3. Wash twice with 1 mL of DRIP Li buffer.

For washing **steps 4** and **5**: Gently add buffer to the tube while it is on the stand. Briefly invert 1–2 times with the stand and remove washing buffers.

4. Wash once with 1 mL of TE + Salt (50 mM NaCl) buffer.
5. Wash once with 1 mL of TE buffer.
6. Spin at 960× *g* for 15 s to pellet any remaining particles, then carefully remove any residual TE buffer.
7. Add 200 μL of elution buffer that has been pre-warmed to 65 ° C.
8. Elute the material from the beads by incubating the tubes at 65 °C for 30 min while shaking at 1400 rpm. In this protocol, we used an Eppendorf ThermoMixer® C for this step.
9. Briefly centrifuge the tubes for 5 s, then immediately place the tubes on the magnetic bead stand to separate the supernatant.
10. Spin down the beads at 16,000× *g* for 1 min at room temperature to remove any beads that may have been contaminated. After this step, carefully transfer the eluate to a new tube.
11. Add 4 μL of Proteinase K (20 mg/mL) and 4 μL of RNaseIn-Plus Inhibitor to the eluate.
12. Incubate the mixture at 42 °C for 30 min.
13. Add five volumes of Buffer PB to one volume of the DNA: RNA solution and mix thoroughly (*see* **Note 19**).
14. Place a MinElute column in the provided 2 mL collection tube in a suitable rack.
15. To bind the DNA to the column, apply the sample to the MinElute column and centrifuge for 1 min. Ensure all traces

of the sample are transferred to the column for maximum recovery.

16. Discard the flow through. Place the MinElute column back into the same tube.
17. To bind the DNA to the column, apply the sample to the MinElute column and centrifuge for 1 min. Ensure all traces of the sample are transferred to the column for maximum recovery.
18. Elute the DNA by adding 12 μL of diluted BufferEB (5 mM Tris–HCl, pH 8.5) to the column.

3.3.5 For DRIP (DNA:RNA Hybrids Immunoprecipitation for the DNA Strand Sequencing)

1. Dilute the DNA:RNA content solution with 5 mM Tris—HCl (pH 8.5) to a final volume of 165 μL.
2. Add 20 μL of 10× RNaseH buffer to the solution.
3. Add 15 μL of RNaseH enzyme (at a concentration of 5 U/μL).
4. Incubate the mixture for 4 h to overnight at 37 °C.
5. Purify the DNA using a MinElute purification column, following the same steps as **"steps 13–17"** in Subheading 3.3.4.
6. Elute the purified DNA in 12 μL of 5 mM Tris–HCl (pH 8.5).

3.3.6 Southern Blot Analysis (Fig. 4)

1. Mix the DRIP products or whole genomic DNA (10 μL) with 15 μL of a solution containing 0.13 N NaOH/3.3 mM EDTA, and incubate at 90 °C for 10 min.
2. Dilute the denatured DRIP products to a total volume of 100 μL with 6× SSC, and spot the solution onto a positively charged nylon membrane (for example, Hybond N+). This can be done using a Bio-Dot apparatus.

   Cross-link the membrane using UV light at an intensity of 125 mJ/cm^2. Pre-hybridize the membrane using ULTRAhyb Ultrasensitive Hybridization Buffer at 42 °C overnight. Following this, add the 5′-^{32}P-labeled DNA or LNA-oligonucleotide probe to the hybridization buffer and seal the hybridization pack. (Our goal was to detect the quantity of R-loop in the telomeric region; thus, we use a DNA probe composed of the sequence 5′--GGGTTAGGGTTAGGGTTA-3′.)
3. Incubate the packed membrane and probe solution overnight at 42 °C (*see* **Note 20**).
4. Wash the membrane three times with a solution of 2× SSC/0.1% SDS for 15 min at either 42 °C or 50 °C. A fraction of the DRIP samples (5%) should be spotted as an input control onto the membrane.

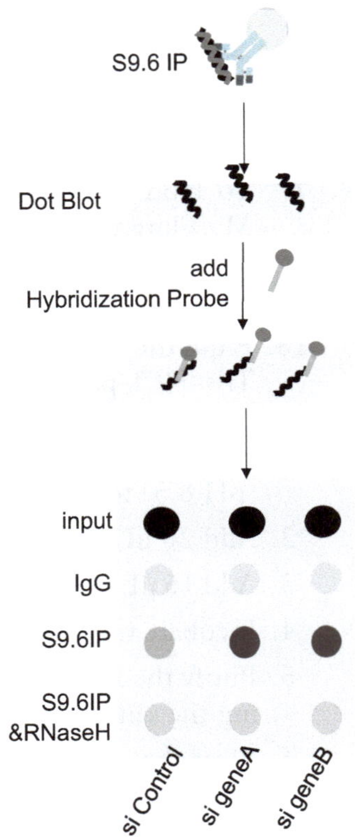

Fig. 4 Workflow of Southern blot analysis. This flowchart details the Southern blot process: Following DRIP using S9.6, dot blot can detect specific regions in the R-loop with the aid of a hybridization probe. In practical experiments, significant analysis of R-loop quantity variations can be achieved by comparing samples with the genes of interest knockdown (genes A, B) to wild type (WT) (S9.6IP), or samples treated with RNaseH, an R-loop resolving factor (S9.6IP + RNaseH)

Detect the signals from the hybridized probes. In this protocol, we used the Typhoon RGB Imager to detect the signals.

4 Notes

1. When preparing a solution of 5% skim milk in PBST (0.1%), be sure to mix and invert the solution for at least 10 min, as the skim milk may not readily dissolve. Skim milk should be prepared fresh for each experiment.
2. For immunostaining of mammalian cells, using an anti-horseradish peroxidase antibody helps decrease background noise by reducing the detection of endogenous peroxidase-like enzymes that are naturally present in the cells.

3. A pellet should be seen in the tube. If no pellet is visible, repeat the centrifugation process.
4. When removing the supernatant, avoid harsh aspiration. Instead, slowly pour off the supernatant multiple times while tilting the tube.
5. In the preparation of 5% skim milk/PBST (0.1%), skim milk can be challenging to dissolve, and 10 min of mixing and inversion can facilitate its dissolution. The solution should be freshly prepared for each experiment.
6. If there are inconsistencies in R-loop volumes among samples, it is advisable to process samples in quadruplicate ($n = 4$) for averaging purposes.
7. When inserting the membrane into the dot blot device, ensure there is no trapped air, which can hinder vacuuming. Rolling the membrane on a rubber plate can help remove excess air and ensure the membrane adheres properly to the rubber sheet.
8. Ensure the screws on the dot blot device are tightened properly. If they are loose, the vacuuming force will be inconsistent across the membrane, potentially leading to sample leakage during the vacuuming process.
9. If there are areas where vacuuming is not happening effectively, refer back to **Notes 6** and **7** to ensure vacuuming is being performed properly.
10. In our laboratory, we use the BIOSAN aspirator (#BS-040108-AAK), which can achieve a vacuum pressure of up to −500 mbar. This level of vacuum pressure is crucial to achieve uniformity and avoid inconsistencies. Both excessive and insufficient vacuum can cause problems.
11. Allow the membrane to dry slightly before placing it in the UV cross-linker. A drying time of approximately 5–10 min is recommended. The membrane should not be wet to the touch and should be slightly wrinkled. Also, make sure the membrane is not floating when placed in the UV cross-linker to ensure consistent cross-linking.
12. Cross-linking can affix the nucleic acids uniformly against the membrane by applying UV from two different directions, one from the top side and then placing it up-side down for another irradiation.
13. During the primary and secondary antibody reactions as well as the washing process, ensure the entire membrane is fully submerged in the solution. If the membrane does not submerge well, consider using a weighting device to guarantee full submersion.

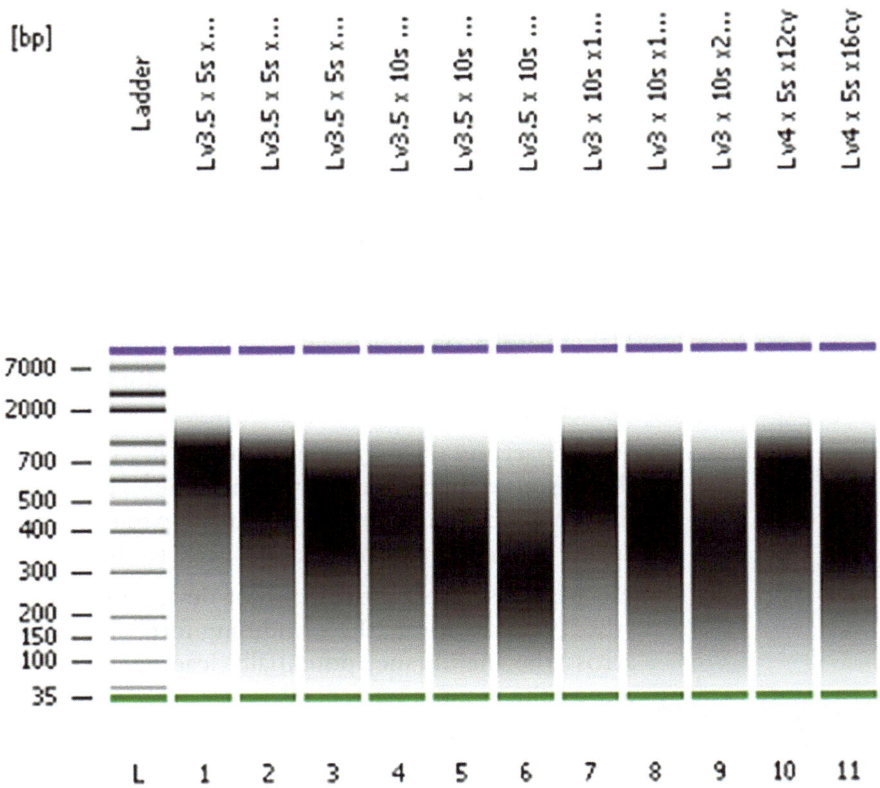

Fig. 5 Sonication Conditions for DNA

14. When shaking the solutions, use a shaker that moves only in a horizontal or vertical direction. Choose a container that matches the size of the membrane to ensure consistent reactions.
15. The results showed different conditions, with the best one being the sonication of DNA in 200–400 bp fragments (*see* Fig. 5).
16. When collecting the beads, use a magnetic bead stand and position the tubes in a rack. Allow the beads to gather on the side of the tube for 15–60 s. After the beads have been collected from washing the lid, ensure to invert the tube twice.
17. This step involves the use of antibodies and beads; therefore, it is recommended to use tubes that promote protein adhesion. Employing suitable tubes can enhance the efficiency of the experiment.
18. All wash steps should be performed at 4 °C. Buffers should be kept cold in a 50-mL tube on ice. Bead collection should be done using a magnetic bead stand, and the supernatant should be removed with an aspirator or pipettor. After resuspending the beads in 1 mL of appropriate wash buffer, the beads should

be collected again. The washing step should be repeated four more times.

19. After adding the pH indicator to Buffer PB, verify that the mixture is yellow. If it appears orange or violet, add 10 μL of 3 M sodium acetate (pH 5.0) and mix well until the solution becomes yellow.
20. The probes used for hybridization and washing temperatures are listed in Table 2. The consensus α-satellite, Alu, and LINE1 probes were hybridized and washed at 42 °C under less stringent conditions. These probes target variations present within each repetitive element's subfamily. The combined consensus probes for Alu and LINE1 recognize approximately 11% and 18% of the human genome, respectively.
21. When adding CPT, the amount of R-loop increases due to the increase in DNA damage. Therefore, in addition to controls such as DNA:RNA hybrids, CPT-treated samples should be prepared to further increase the accuracy of the experiment.

Acknowledgments

We would like to express our profound gratitude, particularly to M. Nakano and Y. Yang for their invaluable assistance in the preparation of this chapter. This work was generously supported by JSPS KAKENHI Grant Numbers JP18H06054, JP19H03159, JP19K16037, and 22H02548, Takeda Science Foundation, Mochida Memorial Foundation for Medical and Pharmaceutical Research, Kowa Life Science Foundation, Astellas Foundation for Research on Metabolic Disorders, MSD Life Science Foundation, The Uehara Memorial Foundation, The Cell Science Research Foundation, and The Sumitomo Foundation.

Table 2
List of control probes and oligonucleotides used for Southern blot hybridization, including washing temperatures

	Source	
DNA probe (T*C*AGGG)3: 42 °C	IDT	5′-GGGTCAGGGTCAGGGTCA-3′
DNA probe (TT*G*GGG)3: 42 °C	IDT	5′-GGGTTGGGGTTGGGGTTG-3′
DNA probe-a-satellite: 42 °C	IDT	5′-CATCACAAAGAAGTTTCTGAGAATGCTTC-3′
DNA probe-Alu: 42 °C	IDT	5′-TCGCGCCACTGCACTCCAGCCTGGGCGACAGAGCGAGACTCCG-3′
DNA probe-LINE1: 42 °C	IDT	5′-TTGGAAAACACTCTGCAGGATATTAT-3′
LNA probe (TTAGGG)3: 42 °C	QIAGEN	5′-GGGT+T + AG + GGT + T + AGGGTTA-3′
LNA probe (TT*G*GGG)3: 50 °C	QIAGEN	5′-GGGT+T + GG + GGT + T + GGGGTTG-3′
LNA probe (CCCTAA)3: 42 °C	QIAGEN	5′-TAACCC+T + AACCC+T + AACCC-3′
LNA probe (CCCT*G*A)3: 55 °C	QIAGEN	5′-TGACCC+T + GACCC+T + GACCC-3′
		(LNA base: +G, +A, +T, +C)

References

1. Aguilera A, García-Muse T (2012) R loops: from transcription byproducts to threats to genome stability. Mol Cell 46(2):115–124
2. Chen PB, Chen HV, Acharya D, Rando OJ, Fazzio TG (2015) R loops regulate promoter-proximal chromatin architecture and cellular differentiation. Nat Struct Mol Biol 22(12): 999–1007
3. Skourti-Stathaki K, Proudfoot NJ (2014) A double-edged sword: R loops as threats to genome integrity and powerful regulators of gene expression. Genes Dev 28(13): 1384–1396
4. Santos-Pereira JM, Aguilera A (2015) R loops: new modulators of genome dynamics and function. Nat Rev Genet 16(10):583–597
5. Yan P, Liu Z, Song M, Wu Z, Xu W, Li K et al (2020) Genome-wide R-loop landscapes during cell differentiation and reprogramming. Cell Rep 32(1):107870
6. Kim A, Wang GG (2021) R-loop and its functions at the regulatory interfaces between transcription and (epi) genome. Biochim Biophys Acta Gene Regul Mech 1864(11–12):194750
7. Nadel J, Athanasiadou R, Lemetre C, Wijetunga NA, Ó Broin P, Sato H et al (2015) RNA: DNA hybrids in the human genome have distinctive nucleotide characteristics, chromatin composition, and transcriptional relationships. Epigenetics Chromatin 8: 1–19
8. Niehrs C, Luke B (2020) Regulatory R-loops as facilitators of gene expression and genome stability. Nat Rev Mol Cell Biol 21(3):167–178
9. Castellano-Pozo M, Santos-Pereira JM, Rondón AG, Barroso S, Andújar E, Pérez-Alegre M et al (2013) R loops are linked to histone H3 S10 phosphorylation and chromatin condensation. Mol Cell 52(4):583–590
10. Ginno PA, Lott PL, Christensen HC, Korf I, Chédin F (2012) R-loop formation is a distinctive characteristic of unmethylated human CpG Island promoters. Mol Cell 45(6):814–825
11. Shiromoto Y, Sakurai M, Minakuchi M, Ariyoshi K, Nishikura K (2021) ADAR1 RNA editing enzyme regulates R-loop formation and genome stability at telomeres in cancer cells. Nat Commun 12(1):1654
12. Bou-Nader C, Bothra A, Garboczi DN, Leppla SH, Zhang J (2022) Structural basis of R-loop recognition by the S9. 6 monoclonal antibody. Nat Commun 13(1):1641

Part II

Computational Approaches to Study RNA-RNA and RNA-DNA Interactions

Chapter 7

In Silico Design of Interacting RNA Sequences Using the MODENA Web Server

Akito Taneda

Abstract

To date, various synthetic RNAs that interact with each other have been proposed to artificially control gene expression. By virtue of efficient secondary structure prediction technologies and powerful optimization algorithms, designing interacting RNAs with functional structures in silico is now possible. In this chapter, we describe the use of a web application to design two interacting RNAs with user-prescribed structural features that are intended to work as a molecular switch to modulate gene expression.

Key words RNA switch, Secondary structure, RNA interaction, Synthetic biology, Multi-objective genetic algorithm

1 Introduction

From the viewpoint of synthetic biology, RNA–RNA interaction is an intriguing and important phenomenon since it provides us a useful way to design two RNAs that specifically bind to each other. Gene expression-controlling systems based on RNAs can be constructed with a riboregulator and a gene expression platform with a terminator for transcriptional control [1] or a ribosome-binding site for translational control [2, 3]. In these systems, for controlling gene expression, a secondary structure is utilized to specify the state of each RNA and to determine the ON/OFF status of gene expression.

Here, we assume the following situations: (1) Before hybridization, both RNAs (a riboregulator and an expression platform) are isolated so that these RNAs fold independently. (2) The two RNAs are hybridized; in this situation, the riboregulator, which has a structure defined in Situation (1), and the expression platform (before hybridization, this expression platform has the structure defined in Situation (1)) coexist. As a result, these two types of RNAs are hybridized, and the gene expression platform is turned

Michiaki Hamada et al. (eds.), *RNA-RNA and RNA-DNA Interactions*, Methods in Molecular Biology, vol. 2949,
https://doi.org/10.1007/978-1-0716-4670-0_7,

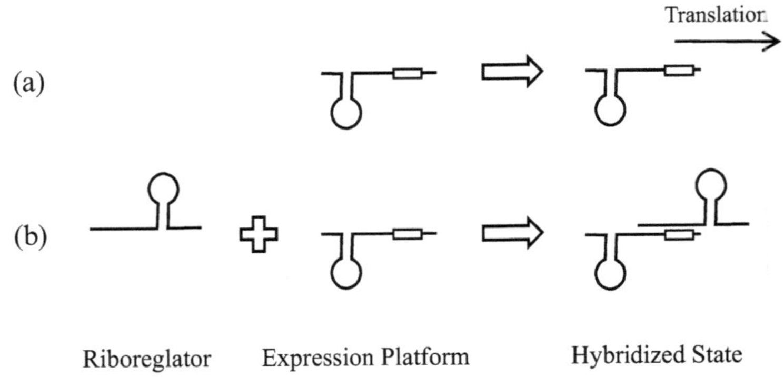

Fig. 1 Two scenarios are assumed in this chapter. (**a**) No riboregulator exists. Translation occurs. (**b**) A riboregulator exists, and the riboregulator and expression platform form a complex (hybridized state). Translation does not occur

ON/OFF in accordance with the formation of the joint secondary structure (a schematic image of these situations is illustrated in Fig. 1).

To design such a set of two RNAs that work in accordance with the user-prescribed scenario, secondary structures (i.e., target structures) encoding a switch function have to be specified by a user. If the user has a set of target structures (those for an isolated riboregulator, an isolated expression platform, and the two hybridized RNAs), e.g., those taken from literature, then the user can use them as target structures to design interacting RNAs. In this chapter, a design example of an OFF switch, using target structures taken from the supplementary data of Ribomaker [3], is described. After its successful design, the user obtains a set of two RNA sequences expected to be folded into the specified structures: the two RNAs form a joint target structure upon hybridization, while forming the independent (isolated) structures before hybridization.

A genetic algorithm (GA) [4] is one of the powerful combinatorial optimization heuristics that have been widely used. In our recent paper [5], we developed the MODENA web server to design two interacting RNAs based on multi-objective GA [6]. Briefly speaking, our GA explores the RNA sequence space to search for a set of two RNA sequences that have user-specified structural features by evolving a set of solutions, where each solution corresponds to a set of a riboregulator and an expression platform. The set of solutions is called the population. In general, we can expect that a larger population yields more optimized design results. In addition, our GA performs sequence sampling based on random numbers to evolve the RNA sequences, and changing the initial random number provides different design results. By specifying these parameters of the GA, we can affect the design results of the MODENA web server. In the rest of this chapter, we describe a step-by-step guide for designing RNA–RNA interaction switches using the MODENA web server.

2 Materials

2.1 Software

Users can design two interacting RNAs using the MODENA web server, so the only software required to perform the design jobs described in the present chapter is a web browser. We confirmed that recent versions of Chrome, Edge, and Firefox web browsers are adequate for designing and browsing the interacting RNA sequences and structures of the design examples described below.

3 Methods

Here, we show a design example of an interacting RNA system composed of two RNAs: a riboregulator and an expression platform. In this example, the expression platform has the Shine–Dalgarno (SD) sequence and the start codon. The secondary structure around the SD sequence and the start codon can modulate the gene expression located downstream of them. If the gene expression platform is sequestered (i.e., a stem structure forms at the region of the expression platform) in response to the riboregulator, the system of the two RNAs works as an OFF switch. On the other hand, if the gene expression platform becomes accessible in response to such a stimulus, the system of the two RNAs corresponds to an ON switch since the gene located downstream of the gene expression platform is expressed when sensing the riboregulator. In this chapter, we design an OFF switch for translationally controlling gene expression as a design example, whereas an ON switch and other expression platforms can also be designed in a similar manner by changing input data such as target structures. The target structures of the OFF switch are taken from the supplementary data of RiboMaker [3]. RNA secondary structures are predicted using the ViennaRNA Package [7] (RNAfold and RNAcofold) to evaluate how well the designed RNA sequences satisfy the specified structural features during the optimization process of interacting RNAs.

3.1 Data Preparation

Target secondary structures are specified in the notation shown in Table 1 (*see* **Note 1**). Sequence constraints written in the IUPAC nucleotide codes are acceptable. An '&' indicates the boundary between the riboregulator and the expression platform.

1. Target structures for the hybridized state of an expression platform and a riboregulator:

```
    ***********************************++++++++++++++++++++
++&********************************************************
************************************************
```

Table 1
The characters used for specifying target structures and structural features in the bracket notation

Characters	Description
`U u . _`	Loop region.
`[ ] ( ) < > { }`	Duplex. Four types of brackets are available.
`?`	Any structure is allowed.
`*`	Any structure is allowed.
`+`	Forming a base pair.
`@`	Forming an intramolecular base pair.
`^`	Forming an intermolecular base pair.
`/ &`	The boundary between two RNAs.

2. A target structure for the isolated state of a riboregulator:

   ``````````
   ******************************************************
   *************************************************
   ``````````

3. A target structure for the isolated state of an expression platform RNA:

   ```
   ***********************************.................
   ....
   ```

4. Sequence constraints for a riboregulator:

   ```
   NNNNNNNNNNNNNNNNNNNNNNNNNNNNNNNNNNNNNNNNNNNNNNNNNNNNNNN
   NNNNNNNNNNNNNNNNNNNCAAAGCCCGCCGAAAGGCGGGCUUUUCUGU
   ```

5. Sequence constraints for an expression platform RNA:

   ```
   NNNNNNNNNNNNNNNNNNNNNNNNNNNNNNNNNNNNNNAGGAGGNNNNNNNN
   NAUG
   ```

6. Specifying prohibited sequence motifs:
   None are specified.

3.2 Submitting Design Job

1. With your browser, browse the home page of the MODENA web server:
   http://rna.eit.hirosaki-u.ac.jp/modena/web/
2. Click the "Design now" button on the web page dedicated to the "RNA–RNA interaction switch (RNAcofold)" (*see* **Note 2**).
3. Copy and paste the target structures and sequence constraints of the OFF switch system to the input form located just below "**step 3**" of the web page of the "RNA–RNA interaction switch (RNAcofold)" (for simplicity, we can skip **steps 1 and 2** of the web page of the "RNA-RNA interaction switch (RNAcofold)"; *see* **Note 3**):

```
***********************************++++++++++++++++++++
++&******************************************************
*************************************************
***********************************.....................
&********************************************************
***********************************************
NNNNNNNNNNNNNNNNNNNNNNNNNNNNNNNNNNNNNNNAGGAGGNNNNNNNNNAUG&N
NNNNNNNNNNNNNNNNNNNNNNNNNNNNNNNNNNNNNNNNNNNNNNNNNNNNNNNNNN
NNNNNNNNNNNNNNNNCAAAGCCCGCCGAAAGGCGGGCUUUUCUGU
```

4. Specify the target range of GC content:

   Set 50% to the central value, and set 10% for the upper bound and lower bound of the target GC content. In this case, RNA sequences whose GC content falls within 40–60% are designed.

5. Specify prohibited polynucleotide tracts:

   Prohibit polynucleotide tracts (e.g., AAAAA) longer than or equal to four nucleotides by selecting "4" in the selection box. Prohibit the polynucleotide tracts of adenine (PolyA), guanine (PolyG), cytosine (PolyC), and uracil (PolyU) by selecting "prohibited" in each corresponding select box.

6. Specify GA parameters:

   Set the population size to 100 (*see* **Note 4**); set the initial random number to the default value (12345; *see* **Note 5**).

7. Click the "Submit this design" button.

   The user's design job starts to run. The screen is automatically moved to the intermediate web page. After the design job is finished, the screen is automatically moved to the design result web page. Usually, in the case of the current version of the MODENA web server, the design job mentioned in this chapter takes a few tens of minutes. If the user has to retrieve the design results later, they can bookmark the intermediate page and browse it later.

3.3 Retrieving and Browsing Design Results

After finishing the design job, browse the design result page. Typically, multiple designed sets of an expression platform and a riboregulator are displayed. Inspect each designed set of two RNA sequences as follows:

1. See the "Str. Dist." column. Zero indicates that the set of designed RNA sequences perfectly satisfies the target structure specification. Otherwise, some nucleotide positions have secondary structures that violate the user-prescribed target structure specification.

2. See the "Ene. Diff." column. This free energy difference equals the following: (the sum of "the energy of the isolated riboregulator" and "that of the isolated expression platform")-("the energy of the hybridized state").
3. See the "Seed" column. The free energy value mentioned in this column is that of the predicted interaction seed. A lower value indicates a better seed. If this value is zero, no seed is found.
4. See the "Designed sequences and structures predicted by RNAcofold and RNAfold" column. The first line in this form indicates the joint secondary structure of the two hybridized RNAs predicted by RNAcofold, followed by the predicted free energy value of the hybridized structure. The secondary structure is displayed in the bracket notation.
5. See the second line in the form that shows the secondary structures of the isolated expression platform and isolated riboregulator, which are predicted by RNAfold; their predicted free energies are indicated at the end of this line.
6. See the third line in the form that provides the set of designed RNA sequences of the expression platform and riboregulator.
7. See the auxiliary information of the designed RNA sequences below the fourth line.

   The auxiliary information is displayed in the following order:

   - Sequences of the detected interaction seed.
   - Predicted stem structure of the detected interaction seed.
   - Detected polynucleotide tracts.
   - The G+C content of the designed RNA sequences falls within the prescribed range or not; specified prohibited motifs are included in the designed RNA sequences or not (in this design example, no prohibited motifs are prescribed as design constraints).
   - Free energy values of unfavored seeds for the homodimers (zero indicates that such a seed is not found).
8. Visualization of predicted structures.

   Click the "Draw by forna" button below the form. Then, secondary structures for the hybridized state (denoted by "sq1&2") and isolated states (denoted by "sq1" and "sq2") are visualized with forna [8] (Fig. 2). The user can move the nucleotides in the figures by dragging them with a mouse.

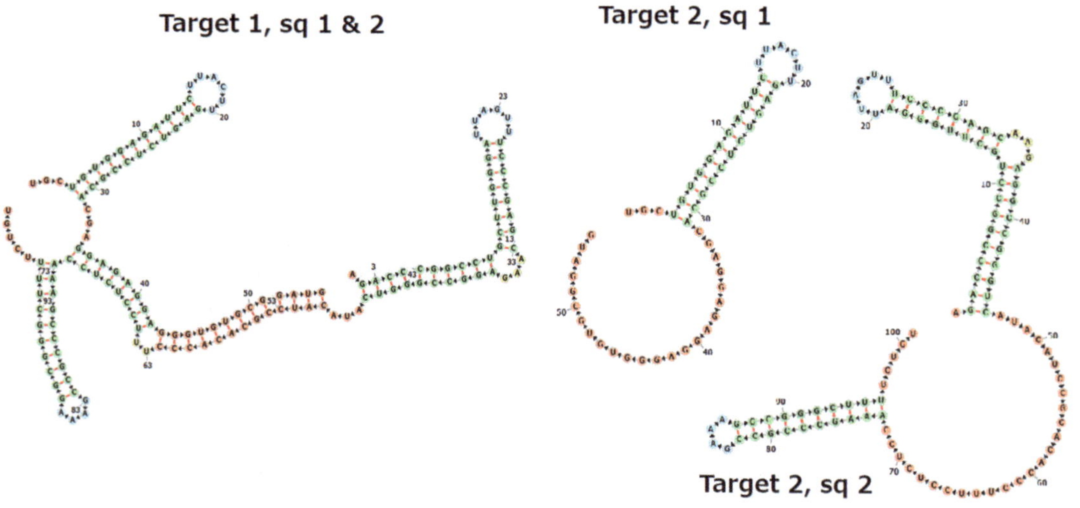

Fig. 2 Example of the secondary structures of the designed OFF switch system

4 Notes

1. *Redesign of the OFF switch.* We can resubmit a design job whose target structures are the predicted secondary structures of the RNA sequences designed in the Subheading 3. Such a redesign usually provides a different set of riboregulator and an expression platform that fold into similar secondary structures obtained via the first design shown in the Subheading 3; this is due to the randomness utilized in the genetic algorithm. This approach may be useful to expand the variation of the designed RNA sequences. An example of the target structures for such a redesign is as follows:

   - Target structures for the hybridized state of an expression platform and a riboregulator:

     ```
     ...(((((((((((......)))))))))))...(((((((((((((((((
     ((((&.((((((((((((((((((......))))))))...))))))))))...
     )))))))))))))..))))))))((((((((((....))))))))))).....
     ```

   - A target structure for the isolated state of a riboregulator:

     ```
     .((((((((((((((((((......))))))))...)))))))))).....
     .....................((((((((((....))))))))))).....
     ```

   - A target structure for the isolated state of an expression platform RNA:

     ```
     ...(((((((((((......)))))))))))................
     ........
     ```

- Sequence constraints for a riboregulator:

NNNNNNNNNNNNNNNNNNNNNNNNNNNNNNNNNNNNNNNNNNNNNN
NNNNNNNNNNNNNNNNNNNNNNNNNCAAAGCCCGCCGAAAGGCGGGCUUUUCUGU

- Sequence constraints for an expression platform RNA:

NNNNNNNNNNNNNNNNNNNNNNNNNNNNNNNNNNNNNNAGGAGGNNNNNNN
NAUG

2. Instead of using RNAcofold and RNAfold, a user can design two interacting RNAs whose secondary structures, predicted using RactIP [9], have the user-prescribed target structures. Click the "Design now" button on the web page dedicated to "RNA–RNA interaction switch (RactIP)." RactIP is the secondary structure prediction method based on integer programming; it can predict joint secondary structures with an external pseudoknot.

3. *Editing target structures in S+ notation*. The target structures used in the Subheading 3 are specified in a somewhat redundant manner. In general, target structures can be dealt with more systematically. For this purpose, the user can utilize the S + notation implemented in the MODENA web server. The target structures used in the Subheading 3 can be written in S + notation as follows (for details on the usage of S+ notation, see Ref. [5] and the help page of the MODENA web server):

```
* 34 A & * 71 B
* 34 U 21 & * 101
A=+++++++++++++++++++++
A=NNNNAGGAGGNNNNNNNNAUG
B=******************************
B=CAAAGCCCGCCGAAAGGCGGGCUUUUCUGU
```

Each integer number following the character (* or U) indicates the nucleotide length value of the character. The characters A and B are user-prescribed structural motifs with sequence constraints. Copy the target structures written in S+ notation and paste them into the input form located just below "**step 1**" of the web page of the "RNA–RNA interaction switch (RNAcofold)." Then click the "Generate target structure (s) and sequence constraints" button. Generated target structures in the bracket notation and sequence constraints are automatically displayed in the input form of "**step 3**" To change the length of the target structures (this changes the length of designed sequences), replace each integer following the character (* or U) with a desired value, then click the "Generate target structure(s) and sequence constraints" button.

4. A larger population size usually provides better designs, but it requires longer run times.
5. Using a different initial random number (e.g., 23456) usually provides different designed results.

References

1. Lucks JB, Qi L, Mutalik VK, Wang D, Arkin AP (2011) Versatile RNA-sensing transcriptional regulators for engineering genetic networks. Proc Natl Acad Sci USA 108(21):8617–8622
2. Rodrigo G, Jaramillo A (2014) RiboMaker: computational design of conformation-based riboregulation. Bioinformatics 30(17): 2508–2510
3. Rodrigo G, Landrain TE, Jaramillo A (2012) De novo automated design of small RNA circuits for engineering synthetic riboregulation in living cells. Proc Natl Acad Sci USA 109(38): 15271–15276
4. Goldberg DE (1989) Genetic algorithms in search, optimization, and machine learning. Addison-Wesley, New York
5. Taneda A, Sato K (2021) A web server for designing molecular switches composed of two interacting RNAs. Int J Mol Sci 22(5):2720
6. Deb K, Pratap A, Agarwal S, Meyarivan T (2002) A fast and elitist multiobjective genetic algorithm: NSGA-II. IEEE Trans Evol Comput 6:182–197
7. Lorenz R, Bernhart SH, Höner Zu Siederdissen C, Tafer H, Flamm C, Stadler PF, Hofacker IL (2011) ViennaRNA Package 2.0. Algorithms Mol Biol 6:26
8. Kerpedjiev P, Hammer S, Hofacker IL (2015) Forna (force-directed RNA): simple and effective online RNA secondary structure diagrams. Bioinformatics 31(20):3377–3379
9. Kato Y, Sato K, Hamada M, Watanabe Y, Asai K, Akutsu T (2010) RactIP: fast and accurate prediction of RNA-RNA interaction using integer programming. Bioinformatics 26(18):i460–i466

Chapter 8

LncRNA:DNA Binding Prediction Using *LongTarget* and Its Revised Version

Jie Lin, Yujian Wen, Hai Zhang, and Hao Zhu

Abstract

Many noncoding RNAs, especially long noncoding RNAs (lncRNAs), can bind to DNA sequences in specific regulatory regions in genomes, thereby regulating gene expression by recruiting epigenomic modification enzymes to their DNA-binding sites (DBSs). Experimental identification of DBSs genome-wide for the abundant lncRNAs in genomes of humans, mice, and other model animals is infeasible, making accurate and fast computational prediction of lncRNA:DNA-binding indispensable. Multiple methods and programs have been developed. This chapter describes *LongTarget* and its variant, *Fasim-LongTarget*, including how to use them step by step to predict DNA-binding domains (DBDs) and DBSs in multiple situations. Both the online and offline versions are described. In addition, we describe how to use the *LongMan* database, which stores orthologous sequences of human and mouse lncRNAs in other mammals, to explore species-specific lncRNAs and lncRNA:DNA binding. We also demonstrate how to evaluate predicted DBDs and DBSs using publicly available data in the UCSC Genome Browser.

Key words LongTarget, Fasim-LongTarget, lncRNA:DNA binding, Triplex, lncRNA, Epigenetic regulation

1 Introduction

Two mechanisms regulate gene transcription. One is the transcription factors (TF)-transcription factor binding sites (TF-DBS) system, and the other is the long noncoding RNAs (lncRNA)-long noncoding RNA–DNA-binding sites (lncRNA-DBS) system. Since TFs are highly conserved in mammalian genomes, a great portion of lncRNAs are highly species-specific [1, 2], and lncRNA genes outnumber protein-coding genes, the importance of transcriptional regulation by the lncRNA-DBS system is drawing increasing attention. Notably, the species specificity of transcriptional regulation is greatly determined by species-specific lncRNAs and their DBSs, and this may explain why transcriptomic differences between humans and macaques and between humans and mice are more significant than previously assumed [3].

Michiaki Hamada et al. (eds.), *RNA-RNA and RNA-DNA Interactions*, Methods in Molecular Biology, vol. 2949,
https://doi.org/10.1007/978-1-0716-4670-0_8,

Just like a small portion of protein-coding genes encode TFs, only a subset of lncRNAs function in the nuclei and have the capability to regulate gene transcription. Also, akin to TFs that use DNA-binding domains (DBDs) with distinct structures (e.g., the homeobox domains and helix–loop–helix domains) to bind to TF-DBSs, lncRNAs follow specific base-pairing rules to make single-strand RNA bind to duplex DNA. Many of these rules are termed Hoogsteen and reverse Hoogsteen rules, as they were initially identified by Karst Hoogsteen [4]. The domains in RNA sequences capable of binding to DNA sequences and the regions in DNA sequences bound by RNA sequences are called triplex-forming oligonucleotides (TFOs) and triplex-targeting sites (TTSs), respectively. Since experimental studies reveal that lncRNAs and epigenome modification enzymes recruited to DBSs show enriched distributions at DBSs [5], a set of densely overlapping TTSs reasonably indicates a DBS, and a set of densely overlapping TFOs reasonably indicates a DBD. In this chapter, we assume this relationship between DBDs and TFOs and between DBSs and TTSs. Besides TFOs, many lncRNAs also contain domains that can bind to proteins, especially DNA and chromatin modification enzymes, which enable them to recruit these proteins to DBSs and epigenetically regulate genes at or near DBSs [6].

Gene expression regulation by lncRNAs plays diverse and important roles. The most studied is genomic imprinting, regulated by lncRNAs such as XIST, H19, Air, and KCNQ1OT1 [7]. Epigenomic regulation of genes by lncRNAs plays critical roles in development and tumorigenesis. It is estimated that the rate of epigenetic changes in cancer cells may be orders of magnitude higher than that of genetic changes [8]. Increasing evidence indicates that gene regulation and dysregulation by lncRNAs critically determine cross-species differences in cancer, cancer metastasis, and cancer immunity [9–11].

In vitro examination of Hoogsteen and reverse Hoogsteen base-pairing rules has been well performed [12], and these rules make it feasible to develop computational methods and programs to predict lncRNA:DNA binding. Predicted DBDs and DBSs have helped address multiple biological questions, for example, the evolution and species specificity of genomic imprinting in mammals [13]. Methods and programs for predicting lncRNA:DNA binding fall into four classes. The first class (e.g., *Triplexator* [14]) identifies triplexes based on the Hoogsteen and reverse Hoogsteen rules by searching RNA sequences against DNA sequences with a few mismatches. This substring search is fast but reports very short triplexes (16–20 bp). Since predicted TTSs are short and barely overlap, false positives are prone to occur, and it is hard to judge which TTSs are true. To overcome this shortcoming, the second class (e.g., *TDF* [15]) of methods integrates statistical methods into the above process to filter out potentially false positives. The third class is *LongTarget*, which features in integrating Hoogsteen and reverse Hoogsteen rules into rulesets, translating the RNA sequence into DNA sequences upon the rulesets, and using a variant of the Smith–Waterman local alignment algorithm to identify aligned sequences [16]. This algorithm reports all local alignments above a predefined threshold instead of the best local alignment. Each ruleset consists of four base-pair rules corresponding to the binding of A, C, G, and T, respectively, and new rulesets can be designed to explore newly identified base-pairing rules. These features enable *LongTarget* to flexibly and reliably identify not only the short TFOs and TTSs but also the quite long DBDs and DBSs. Many *LongTarget*-identified DBSs exceed 100 bp. Nevertheless, the shortcoming of *LongTarget* is also apparent—the local alignment that identifies all aligned sequences makes it highly time-consuming. To overcome this shortcoming, a revised version called *Fasim-LongTarget* has been developed [17]. *Fasim-LongTarget* is based on a revision of the local alignment algorithm, takes much less time, and reports largely the same numbers of DBSs (but fewer TTSs). Instead of using reported Hoogsteen and reverse Hoogsteeen rules, the fourth class (e.g., *TriplexAligner* [18]) uses the RNA and DNA sequences in experimentally identified triplexes in some cell lines to train a probabilistic model and uses the trained model to predict DBSs. The trained model may contain multiple matrices that describe interactions between nucleotides. Training a model using experimentally identified DBD and DBS sequences is the merit of this method, but it is also a risk because not all experimentally identified DBD and DBS sequences are true. In addition, DBSs in many cell types may not be available, which may greatly influence the performance of lncRNA:DNA-binding prediction.

Here, we describe the use of *LongTarget* and *Fasim-LongTarget*. Both are accessible on the web server and can also be executed locally on the user's computer after compilation.

2 Materials

2.1 The LongTarget Website and Program

The online *LongTarget* is available at http://www.gaemons.net/LongTarget. The source code written in C++ can be obtained from https://github.com/LongTarget/.

2.2 Fasim-LongTarget

Fasim-LongTarget is also available on the *LongTarget* website, and the C++ code can be found at https://github.com/LongTarget/Fasim-LongTarget.

2.3 LncRNA and DNA Sequences

The *LongMan* database (http://www.gaemons.net/LongManSearch), which is affiliated to *LongTarget* and can also be used independently, contains 13562 GENCODE-annotated human lncRNAs, 10481 GENCODE-annotated mouse lncRNAs, the orthologous sequences of the 13562 human lncRNAs in 16 mammalian genomes (including primates, rodents, and other mammals) [19], and the orthologous sequences of the 10481 mouse lncRNAs in four mammalian genomes (including three rodents and the rabbit). We searched the orthologous sequences of each exon of an lncRNA using the *Infernal* program [20, 21]. The *LongMan* database also contains the DNA sequences of multiple species from the Ensembl (https://www.ensembl.org/), UCSC (http://genome.ucsc.edu/), and NCBI (https://www.ncbi.nlm.nih.gov/) websites. Since about one-third of human and mouse lncRNAs are primate- and rodent-specific, the orthologues of human and mouse lncRNAs in other mammals are valuable for disease and evolution studies. They help identify what human lncRNAs are human, simian, and primate-specific and what mouse lncRNAs are mouse- and rodent-specific.

2.4 The UCSC Genome Browser

We use the UCSC Genome Browser (http://genome.ucsc.edu/) to visualize the distribution of predicted DBSs in a genomic context with rich information. The genomic context includes annotated genes, transposable elements, ENCODE-annotated cis-regulatory elements, and epigenomic signals in multiple cell lines. The predicted DBSs in a genomic region, a chromosome, or even genome-wide, are formatted into track files of the UCSC Genome Browser, and the track files are uploaded to the UCSC Genome Browser as custom tracks.

3 Methods

3.1 Predicting lncRNA:DNA Binding Using Online LongTarget and Fasim-LongTarget

3.1.1 Predicting lncRNA: DNA Binding at a Given Genomic Locus

Predicting lncRNA:DNA binding on a given genomic locus has two situations—(a) predicting the DBSs of one lncRNA and (b) predicting the DBSs of all lncRNAs in a species. We use the online version to demonstrate the first situation step by step, taking the human lncRNA PTENP1-AS and its potential target gene *PTEN* as an example.

1) **Input DNA Sequence**. In the UCSC Genome Browser, *PTEN* is in the genomic region chr10:89,623,195-89,728,532 (coordinates in hg19). Thus, on the *LongTarget* website, choose or fill in the boxes "Species," "Chr," "From," and "To," then press the "Retrieve" button, which fetches the DNA sequence of the chosen genomic region from the hg19 stored in the website, and insert it into the input box (Fig. 1).

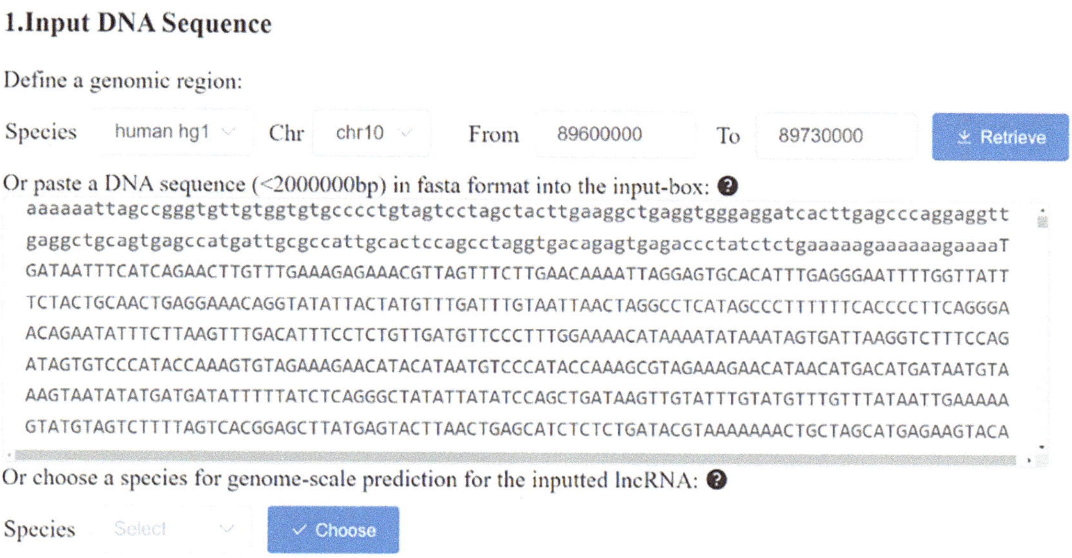

Fig. 1 Choose a DNA sequence from the genome sequences stored in the *LongTarget* website

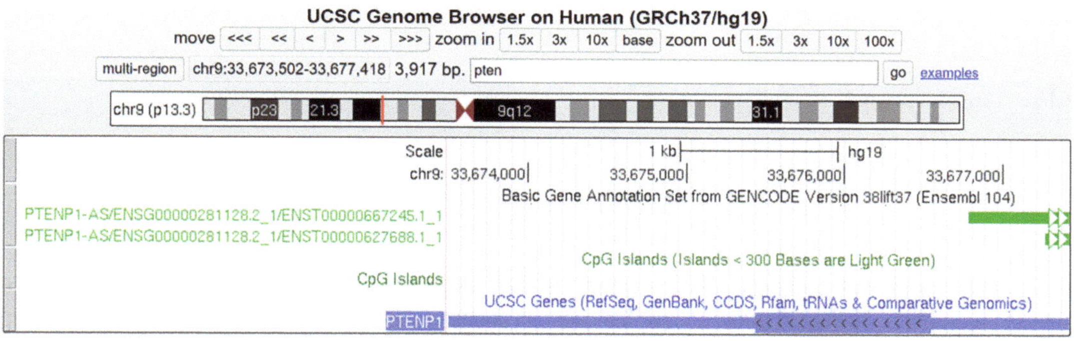

Fig. 2 Identify an lncRNA transcript and obtain its sequence from the UCSC Genome Browser

2) **Input LncRNA Sequence**. We obtain the human lncRNA PTENP1-AS by searching the keyword "pten" in the UCSC Genome Browser (hg19). The search makes the genome browser locate the genes *PTEN* and *PTEN-AS1* (Fig. 2). We choose the transcript ENST00000627688.1_1 by clicking the **ENST00000627688.1_1** to enter the ENST00000627688.1_1 (PTENP1-AS) webpage. Then, click the **Predicted mRNA** under the title **Sequences**, copy the shown sequence (including the title line ">ENST00000627688.1_1"), and paste the sequence into the **Input LncRNA Sequence** box in the *LongTarget* webpage (Fig. 3).

3) Alternatively, the user can directly paste a DNA sequence into the **Input DNA Sequence** box and an lncRNA sequence into the **Input LncRNA Sequence** box. These sequences should

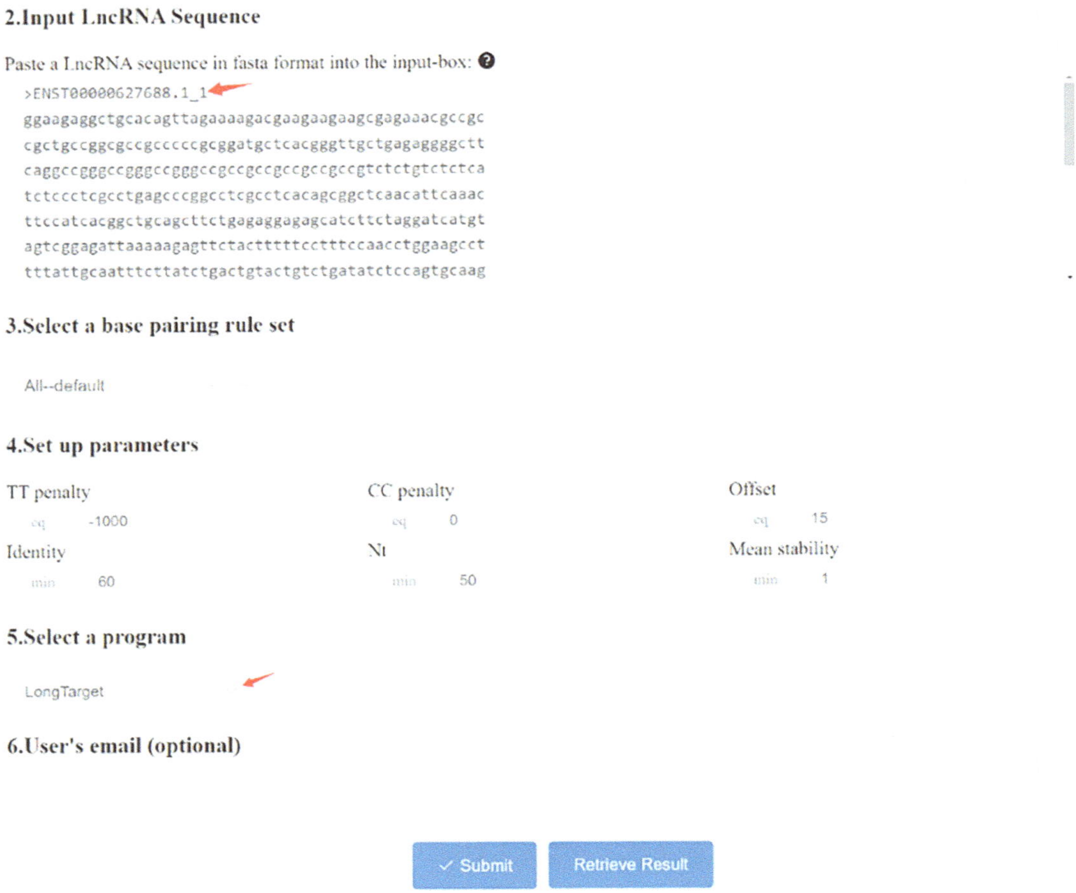

Fig. 3 Input an lncRNA sequence into the **Input LncRNA Sequence** box in the *LongTarget* webpage and choose parameters. As the red arrow shows, the user can choose *LongTarget* or *Fasim-LongTarget* at this step

be in the following formats, and note that the title lines and sequences should contain no hyphens except for the hyphen between the start and end addresses of the DNA sequence.

```
>hg19|chr1|12345678-12346678
ACTGACTG...
> PTENP1_AS
GGACCGCAG...
```

4) By default, *LongTarget* and *Fasim-LongTarget* examine all built-in base-pairing rulesets (Fig. 3). The user can also choose a particular ruleset to predict lncRNA:DNA binding determined by this selected base-pairing ruleset. Parameters in the default setting are suitable for most situations, and these parameters can work with both *LongTarget* and *Fasim-LongTarget*. The most critical parameter is "Nt," which is a length threshold for predicted DBSs. The second critical parameter is "Offset," which determines whether two adjacent triplexes are generated by the same DBD or by two different DBDs. A small "Nt" (e.g., 40) should work with a small "Offset" (e.g., 10), which

makes *LongTarget* report many DBSs (most of which may be short). Instead, a large "Nt" (e.g., 80) should work with a large "Offset" (e.g., 20), which makes *LongTarget* report longer but fewer DBSs. Given the same Offset value, long DBSs are more reliable than short ones. *Fasim-LongTarget* is much faster than *LongTarget* while predicting largely the same DBSs with comparable accuracy.

5) Leaving an email address, which makes the results emailed automatically, is preferable or even necessary because large-scale prediction may take hours and even days. After submitting a task by clicking the "Submit" button, a job ID is generated that can be used to retrieve the results manually by clicking the "Retrieve Result" button (which takes the user to the retrieve result webpage).

6) *LongTarget* and *Fasim-LongTarget* generate the same output files: (a) a *TFOclass1* file contains the DBSs of DBD1, (b) a *TFOclass2* file contains the DBSs of DBD2, (c) a *TFOsorted* file contains details of all TFOs and TTSs, (d) a *BestRule* file indicates the dominating base-pairing ruleset of all triplexes, and (e) two PDF files show the distributions of DBD1's DBSs and DBD2's DBSs in a UCSC Genome Browser-style context. *LongTarget* predicts multiple DBDs, especially if the lncRNA is very long (*Fasim-LongTarget* predicts fewer ones), but not all DBDs have an equal likelihood to be true. Thus, an algorithm is designed to sort DBDs upon their DBS numbers, and DBD1 and DBD2 are the best and second-best DBDs which have the most and the second most DBSs.

7) To view DBSs in a rich genomic context, the user can upload the *TFOclass1* and *TFOclass2* files to the UCSC Genome Browser as two custom tracks (using the "add custom tracks" function in the browser). This allows predicted DBSs to be displayed together with other genomic and epigenomc information, including genes, CpG Islands, ENC DNA Methylation, and ENC Histone Modification tracks.

3.1.2 Predicting lncRNA: DNA-Binding Genome-Wide

The online versions of *LongTarget* and *Fasim-LongTarget* allow only genome-wide DBS prediction of one lncRNA in the promoter regions of all Ensembl-annotated transcripts. To perform this genome-wide prediction, the user should choose the "Or choose a species for genome-scale prediction for the inputted lncRNA" function on the *LongTarget* webpage (Figs. 1 and 4). Promoter regions are defined as −3500~+1500 upstream/downstream of the transcription start sites (TSS) of annotated transcripts. *LongTarget* takes about 10 to 15 days (depending on the lncRNA) to scan all transcripts in the human genome build hg38, but *Fasim-LongTarget* takes just about two days. This genome-wide prediction returns an Excel file that contains coordinates and length of all DBSs, as

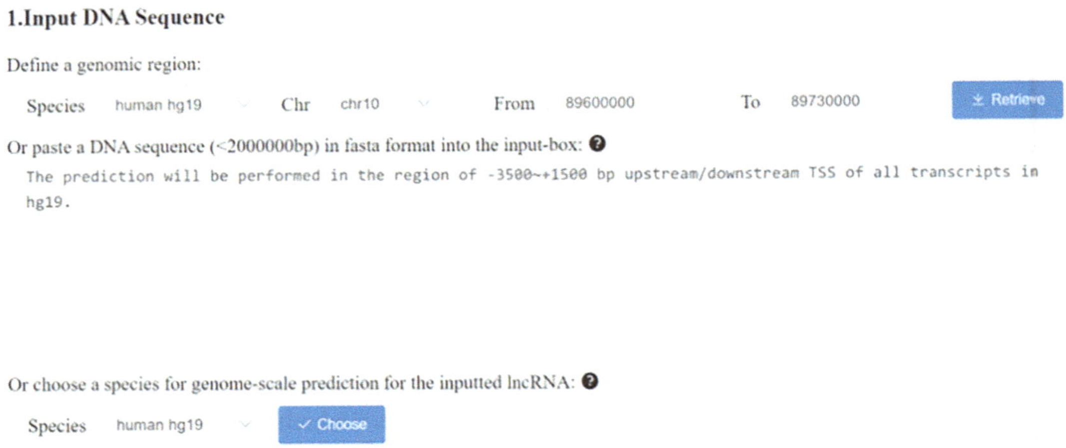

Fig. 4 Making genome-wide lncRNA:DNA-binding prediction upon Ensembl-annotated transcripts and using genome sequences stored in the *LongTarget* website

bs_chr	bs_start	bs_end	transcript_name	gene_name	gene_EnsemblID	gene_type	DBS_strength
chr1	181080	181548	FO538757.3-201	FO538757.3	ENSG00000279928	protein_coding	5145
chr1	195270	195365	FO538757.2-201,202	FO538757.2	ENSG00000279457	protein_coding	358
chr1	199730	200041	FO538757.2-203	FO538757.2	ENSG00000279457	protein_coding	1476
chr1	188367	188524	MIR6859-2-201	MIR6859-2	ENSG00000273874	miRNA	715
chr1	199730	200041	FO538757.1-201	FO538757.1	ENSG00000275135	miRNA	2007
chr1	267350	267444	AP006222.2-003	AP006222.2	ENSG00000228463	lincRNA	539
chr1	267350	267444	AP006222.2-002	AP006222.2	ENSG00000228463	lincRNA	477
chr1	924986	925175	SAMD11-011	SAMD11	ENSG00000187634	protein_coding	862
chr1	924986	925175	SAMD11-003	SAMD11	ENSG00000187634	protein_coding	862
chr1	924986	925175	SAMD11-010	SAMD11	ENSG00000187634	protein_coding	1008

Fig. 5 A schematic output of the genome-wide lncRNA:DNA-binding prediction. Each line indicates a DBS in the promoter region of an annotated transcript

well as gene symbol and Ensembl IDs of target transcripts (Fig. 5). DBS strength is defined as the "size" of a DBS (the length times the height), which is larger than DBS binding affinity (the length times the identity; 'identity' is the percentage of paired RNA:DNA nucleotides in a triplex), which is another measure we use to evaluate lncRNA:DNA binding. By default, DBSs with a size>300 are outputted into the Excel file.

3.2 Running LongTarget *and* Fasim-LongTarget *on the User's Computer*

This section describes running the offline versions of *LongTarget* and *Fasim-LongTarget*. The C++ codes of *LongTarget* and *Fasim-LongTarget* are deposited on the GitHub website, which can be conveniently downloaded and compiled on the user's computer. We recommend this approach for the second type of genome-wide prediction (e.g., predicting all lncRNAs' DBSs in a genomic region) and for true genome-wide prediction (not restricting the DNA sequences to the promoter regions).

3.2.1 Predicting all lncRNAs' DBSs in a Genomic Region

We predict the DBSs of the 13562 GENCODE-annotated human lncRNAs in the genomic region of the *CDKN2A/2B* genes. To do this, we obtain the sequences of the 13562 human lncRNAs from the *LongMan* database, store the sequence files in the "Gencode_lncRNA/" directory, (one file contains one sequence), and create this directory if it does not exist on the user's computer.

1) First, the user should download the *LongTarget* ZIP file from the GitHub website and unzip the file using the "unzip" command. The C++ code needs to be compiled to generate the executable *LongTarget* program, typically using the following command under the Linux system:

```
$ g++ longtarget.cpp -O -msse2 -o LongTarget
```

2) After compilation, the executable *LongTarget* file will be produced in the current directory. If the user chooses to use the *Fasim-LongTarget* program, the compilation command is as follows:

```
$ g++ Fasim-LongTarget.cpp ssw_cpp.cpp sswNew.cpp -O
-msse2 -o fasim-LongTarget
```

3) To check the parameters for running *LongTarget*, the user can type this command:

```
./LongTarget -help
```

4) Important command-line arguments of the two programs include "f1," "f2," "O," "ds," and "lg" (Fig. 6). "f1" and "f2" specify the DNA and RNA files; "O" is the directory into which the result files are written; "lg" and "ds" correspond to the "Nt" and "Offset" parameters mentioned above.

A

Options	Parameters	Functions
f1	DNA sequence file	A string, indicate the DNA sequence file name.
f2	RNA sequence file	A string, indicate the RNA sequence file name.
r	rules	An integer, indicate base-pairing rules, "0" indicates all rules.
O	Output path	A string, indicate the directory into which the results are outputted.
c	Cutlength	An integer, indicate the length of each segment, the default value is 5000.
i	identity	An integer, indicate the criterion of alignment output, the default value is 60.
S	stability	A floating point, indicate the criterion of base-pairing, the default value is 1.0.
ni	ntmin	An integer, indicate the min length of triplexes, the default value is 20.
na	ntmax	An integer, indicate the max length of triplexes, the default is 100000 but is rarely used
pc	penaltyC	An integer, indicate penalty, the default value is 0.
pt	penaltyT	An integer, indicate penalty, the default value is -1000.
ds	c_dd	An integer, indicate the distance between TFOs, the default value is 15.
lg	c_length	An integer, indicate the min length of triplexes, the default value is 50.

B

```
import os

for rnaFile in os.listdir('Gencode_lncRNA/'):
    #13562 sequence files are stored in Gencode_lncRNA/
    #CNKN2A2B.fa is the DNA sequence of CNKN2A/2B locus
    cmd = './fasim-LongTarget -f1 CNKN2A2B.fa -f2 Gencode_lncRNA/%s -O results/' % rnaFile
    os.system(cmd)
```

Fig. 6 The command-line arguments and an exemplary script for running *LongTarget*. (**a**) The command-line arguments. (**b**) An exemplary Python program for using *Fasim-LongTarget* to predict all lncRNAs in a genomic region

The arguments for running *Fasim-LongTarget* are similar to the above ones. A simple example is:

```
$ ./LongTarget -f1 DNA.fa -f2 RNA.fa -O results/
```

5) To predict the DBSs of the 13562 human lncRNAs in the *CDKN2A/2B* genes (chr9:21974000-22012500, hg19), a Python script with just a few lines is needed to implement a loop to examine all lncRNAs. In this script, *LongTarget* or *Fasim-LongTarget* is called (Fig. 6). To predict DBSs of all lncRNAs in a genomic region of some other species, the user should just slightly revise the small script, organizing lncRNAs into files in the "Gencode_lncRNA/" directory and creating the directory for saving result files.
6) The *TFOsorted, TFOclass1*, and *TFOclass2* files for each lncRNA are generated in the given directory that contains other results. The user can upload multiple *TFOclass1* files to the UCSC Genome Browser as custom tracks using the "add custom tracks" button on the UCSC Genome Browser to visualize DBS distributions (Fig. 7). As mentioned above, these tracks can be displayed with other built-in tracks, such as UCSC Genes, CpG Islands, ENC DNA Methylation, and ENC Histone Modification. Note that the genome version of the input DNA sequence and the UCSC Genome Browser should be the same (e.g., both are hg19 or hg38)

3.2.2 Predicting Genome- and Chromosome-wide DBSs of a lncRNA

The time consumption of *LongTarget* makes the truly genome-wide and chromosome-wide (i.e., not just in promoter regions of genes) prediction infeasible. *Fasim-LongTarget* enables this type of prediction. For example, the user can predict PTENP1-AS's DBSs on the whole chr21 chromosome of hg19 using the command:

```
$ ./fasim-LongTarget -f1 chr21.fa -f2 PTENP1AS.fa -O results/
```

Running takes only 11 hours (using a Xeon(R) E7-4830 v3, 2.10 GHz). The prediction creates *TFOsorted* and *TFOclass1 and TFOclass2* files that contain the predicted DBSs of TFO1 and TFO2 in PTENP1-AS on chromosome 21. To perform genome-wide predictions, simply apply the command to all chromosomes (e.g., chr1.fa, chr2.fa, and chr22.fa). To conduct genome-wide predictions for other species, apply the command to all chromosomes of those species.

3.3 Exploring the LongMan Database for Predicting lncRNA:DNA Binding

As mentioned, *LongMan* stores the orthologous sequences of human and mouse lncRNAs across multiple other mammalian genomes. We utilize the orthologue of human *H19* in marmosets and the DNA sequence of the *IGF2* gene in the marmoset genome

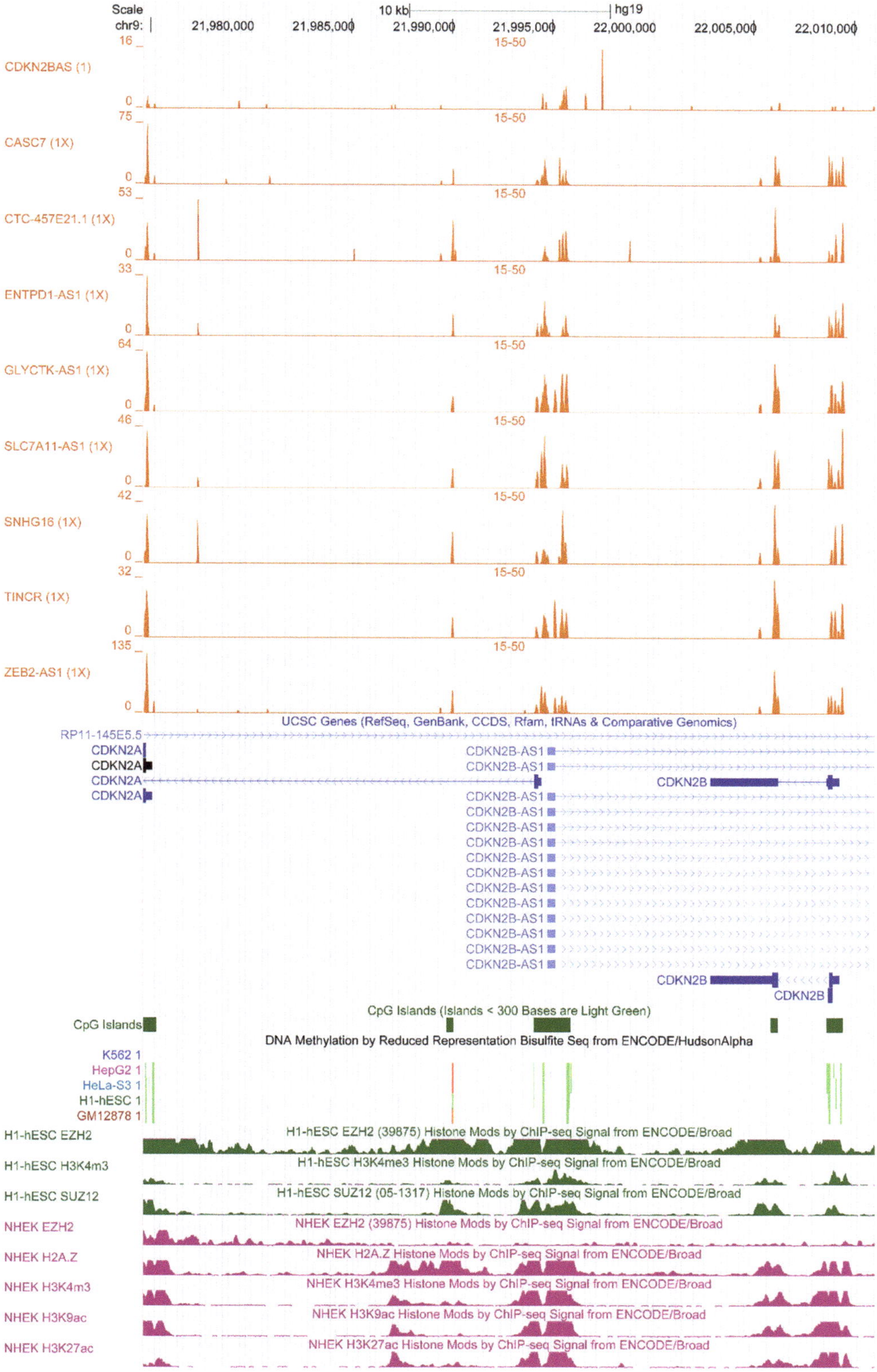

Fig. 7 The DBS distributions of all human lncRNAs (only some are displayed) in the human CDKN2A/2B locus

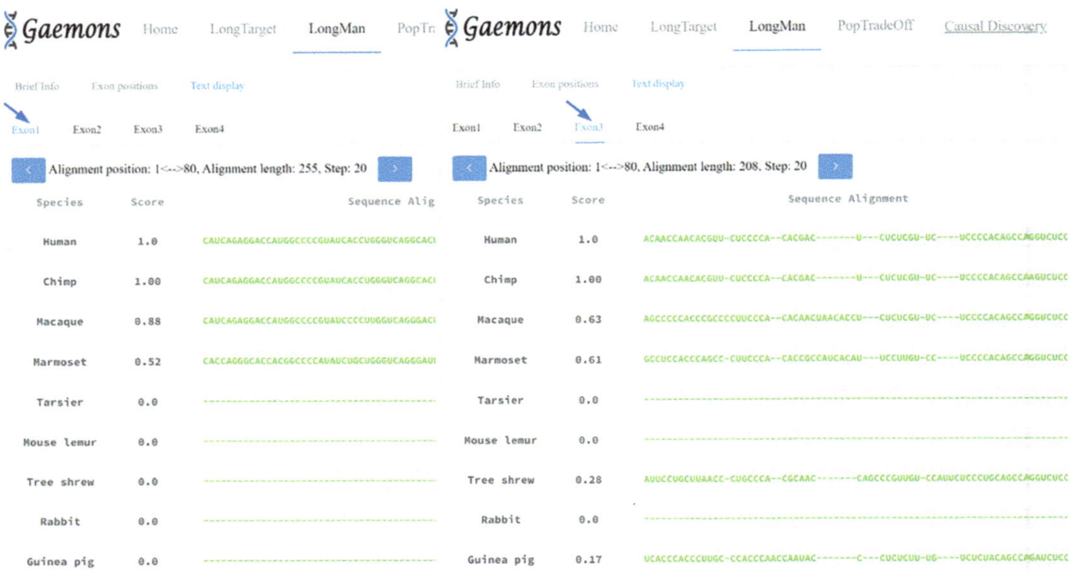

Fig. 8 Human H19 comprises exons originating in different mammals

to illustrate how to explore the *LongMan* database for predicting lncRNA:DNA binding in other species. This database and its DBS prediction of species-specific lncRNAs are important for two reasons. First, the species specificity of lncRNAs has garnered increasing attention. A key question is which human lncRNAs are specific to primates, simians, and humans, and which mouse lncRNAs are specific to rodents and mice. Since the orthologous sequences of human and mouse lncRNAs have been identified and stored in the *LongMan* database, users can search them by entering an Ensembl ID, HGNC gene symbol, or gene name. Taking the human lncRNA H19 as an example, it is noteworthy that this lncRNA, known to be highly conserved in mammals, actually comprises exons derived from different mammals (Fig. 8). Therefore, the DBDs predicted in specific exons may indicate when H19 acquired its epigenetic regulatory function. Second, H19 regulates the genomic imprinting of *IGF2* in both humans and mice. Genomic imprinting, particularly the imprinted expression of *IGF2*, is vital for embryonic development, but the evolutionary origins of this regulation are much less understood.

To examine whether the regulation of *IGF2* by H19 possibly exists in marmosets, the user can use the sequence of human H19's orthologue in marmosets and the sequence of the *IGF2* gene in marmosets to make a prediction. Since the Ensembl ID of human H19 is ENSG00000130600.11, we let human H19's orthologue in marmoset be named ENSG00000130600.11-marmoset in

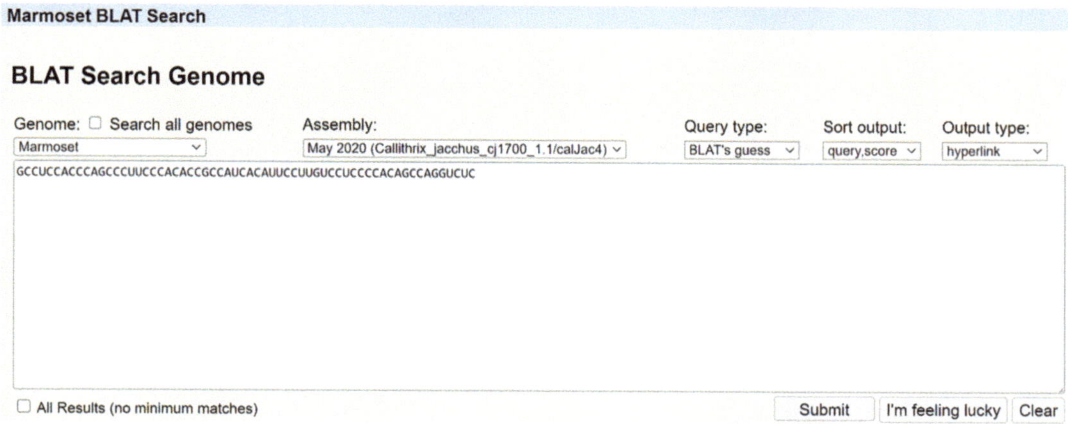

Fig. 9 Search the marmoset genome using the orthologous sequences of one or multiple H19 exons

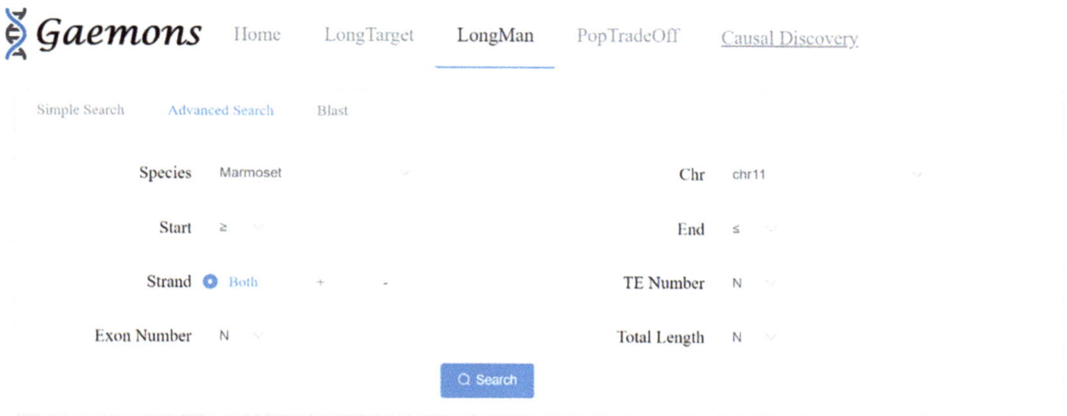

Fig. 10 Search the target orthologous sequence in marmoset using the Advanced Search function

LongMan. We currently use the UCSC Genome Browser to identify in which chromosome ENSG00000130600.11-marmoset is located. We use the *BLAT* function in the UCSC Genome Browser to search the marmoset genome with the orthologous sequences of one or multiple H19 exons (it is better to choose long exons) (Fig. 9). After confirming that the sequence of ENSG00000130600.11-marmoset is on marmoset chromosome 11, we use the **Advanced Search** function in the *LongMan* database to obtain ENSG00000130600.11-marmoset (Fig. 10). In the webpage that shows the search results (Fig. 11), identify ENSG00000130600.11-marmoset and double-click this record to open the page showing the sequence of ENSG00000130600.11-marmoset (Fig. 12).

There are two ways to obtain the sequence of the annotated *IGF2* gene in the marmoset genome, and which one to use depends

smu_id	species	HGNC gene symbol	gene_name(alias_name)	Chr	strand	start	end
52997	Marmoset	ENSG00000235027.1_Marmoset		chr11	-	128675015	128677077
53000	Marmoset	ENSG00000229512.1_Marmoset		chr11	+	128674452	128674495
53005	Marmoset	ENSG00000230834.1_Marmoset		chr11	+	128652938	128654340
53012	Marmoset	ENSG00000249086.1_Marmoset		chr11_GL285518_random	-	27905	28070
53016	Marmoset	ENSG00000226416.1_Marmoset		chr11	+	128455231	128461224
53023	Marmoset	ENSG00000232987.1_Marmoset		chr11	-	128451939	128454054
53030	Marmoset	ENSG00000130600.11_Marmoset		chr11	+	128443118	128450321
53039	Marmoset	ENSG00000240801.1_Marmoset		chr11	+	128307710	128308546
53043	Marmoset	ENSG00000099869.6_Marmoset		chr11	-	128295400	128295822
53054	Marmoset	ENSG00000230483.1_Marmoset		chr11	-	128040820	128044915

Total 646 10/page < 1 2 3 4 5 6 ••• 65 > Go to 3

Fig. 11 Identify ENSG00000130600.11-marmoset in the search results page and double-click this line to open the page showing its sequence

on the version. The sequence of WUGSC 3.2/calJac3 (released in March 2009) is available on the *LongTarget* website, while the sequence of Callithrix_jacchus_cj1700_1.1/cal (released in May 2020) is not. In both cases, it is necessary to use the UCSC Genome Browser to find the address of *IGF2* (Fig. 13). To predict the DBS of H19 in *IGF2* in the marmoset genome WUGSC 3.2/calJac3, the user can use the *LongTarget* website by selecting the genomic region. To predict the DBS of H19 in *IGF2* in the marmoset genome Callithrix_jacchus_cj1700_1.1/cal, the user should fetch, copy, and paste the genomic sequence of *IGF2* into the **Input DNA Sequence** box, or the user can use the standalone version of *LongTarget* or *Fasim-LongTarget*.

Fig. 12 The sequence of ENSG00000130600.11-marmoset in the *LongMan* database

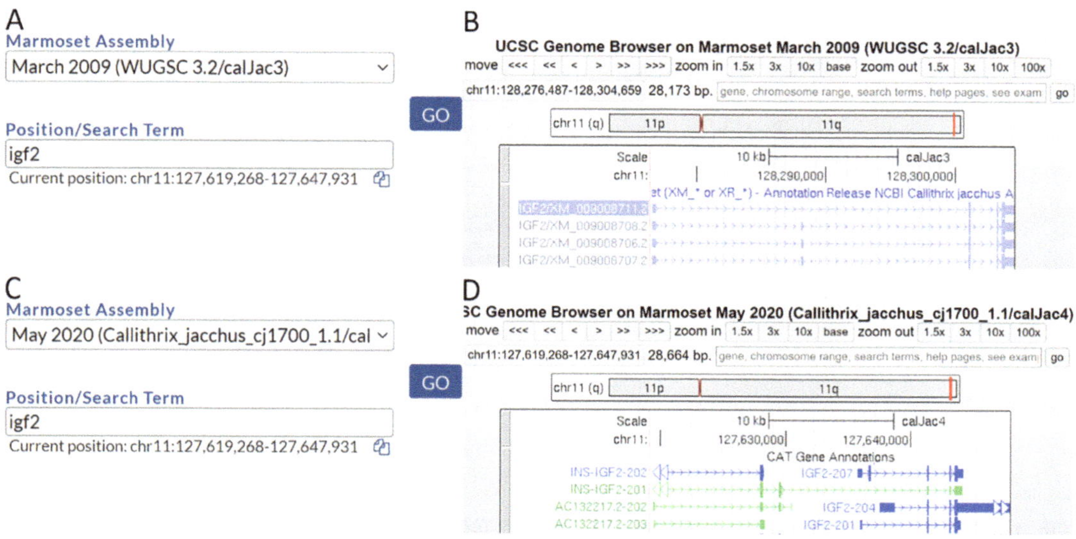

Fig. 13 Obtaining the sequence of the *IGF2* gene in the marmoset genome. (AB) Search the *IGF2* gene in the marmoset genome WUGSC 3.2/calJac3. (CD) Search the *IGF2* gene in the marmoset genome Callithrix_jacchus_cj1700_1.1/cal

4 Evaluating Predicted DBSs

Predicted DBSs can be evaluated based on three categories of data: experimentally identified DNA methylation and histone modification signals in cell lines (Fig. 14), experimentally identified lncRNAs' DBSs in certain cell lines (Fig. 15), and annotated cis-regulatory elements in the human genome (Fig. 16). DBSs at DNA methylation and histone modification signals or annotated cis-regulatory elements are biologically plausible. In addition, theoretically, long DBSs (or DBSs with larger strength or binding affinity) are more reliable than short ones. Further, if a set of DBSs is predominantly determined by DBD1 instead of by multiple different DBDs, these DBSs are also more reliable because they are likely predicted upon one ruleset instead of different rulesets.

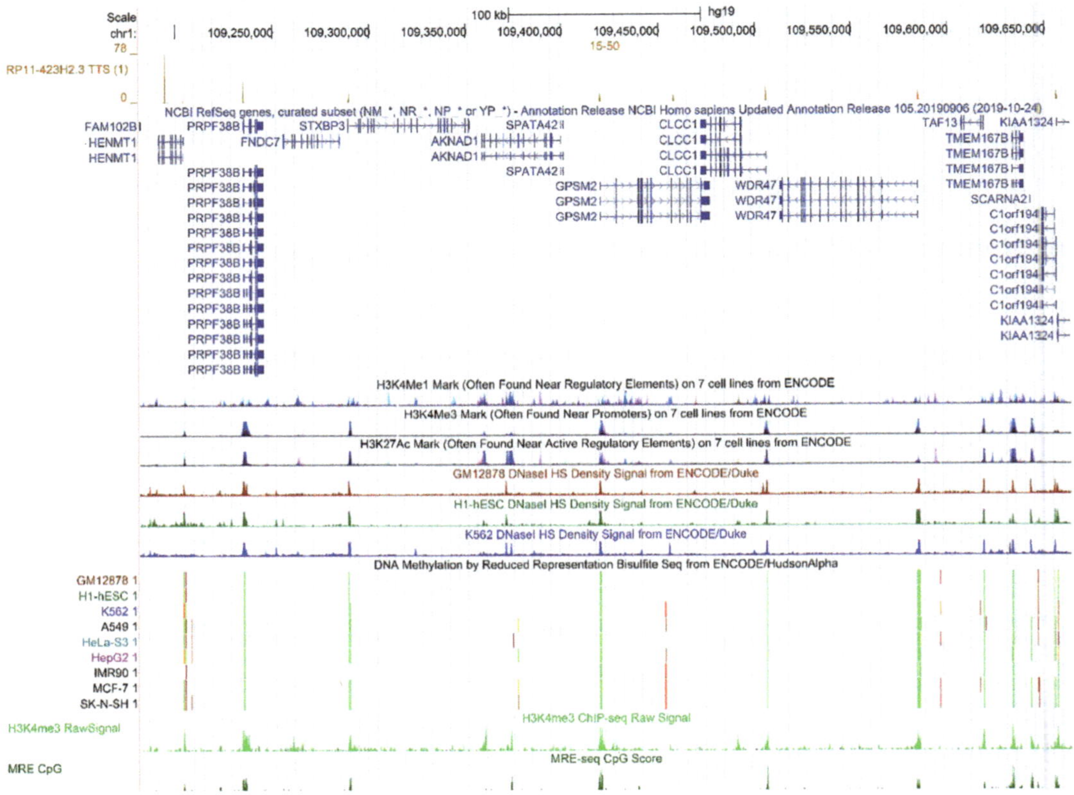

Fig. 14 The DBS distribution of the lncRNA RP11-423H2.3 (AC106795.2/ENSG00000249684.6) in a genomic region. The *TFOclass1* file, which contains the genome-wide DBSs of RP11-423H2.3 in the promoter regions of annotated transcripts, is uploaded to the UCSC Genome Browser as a custom track (the top track). The following tracks are UCSC gene annotation, histone modification signals in some cell lines, DNA methylation signals in some cell lines, and CpG Islands. These epigenomic modification signals in cell lines are stored in the UCSC Genome Browser

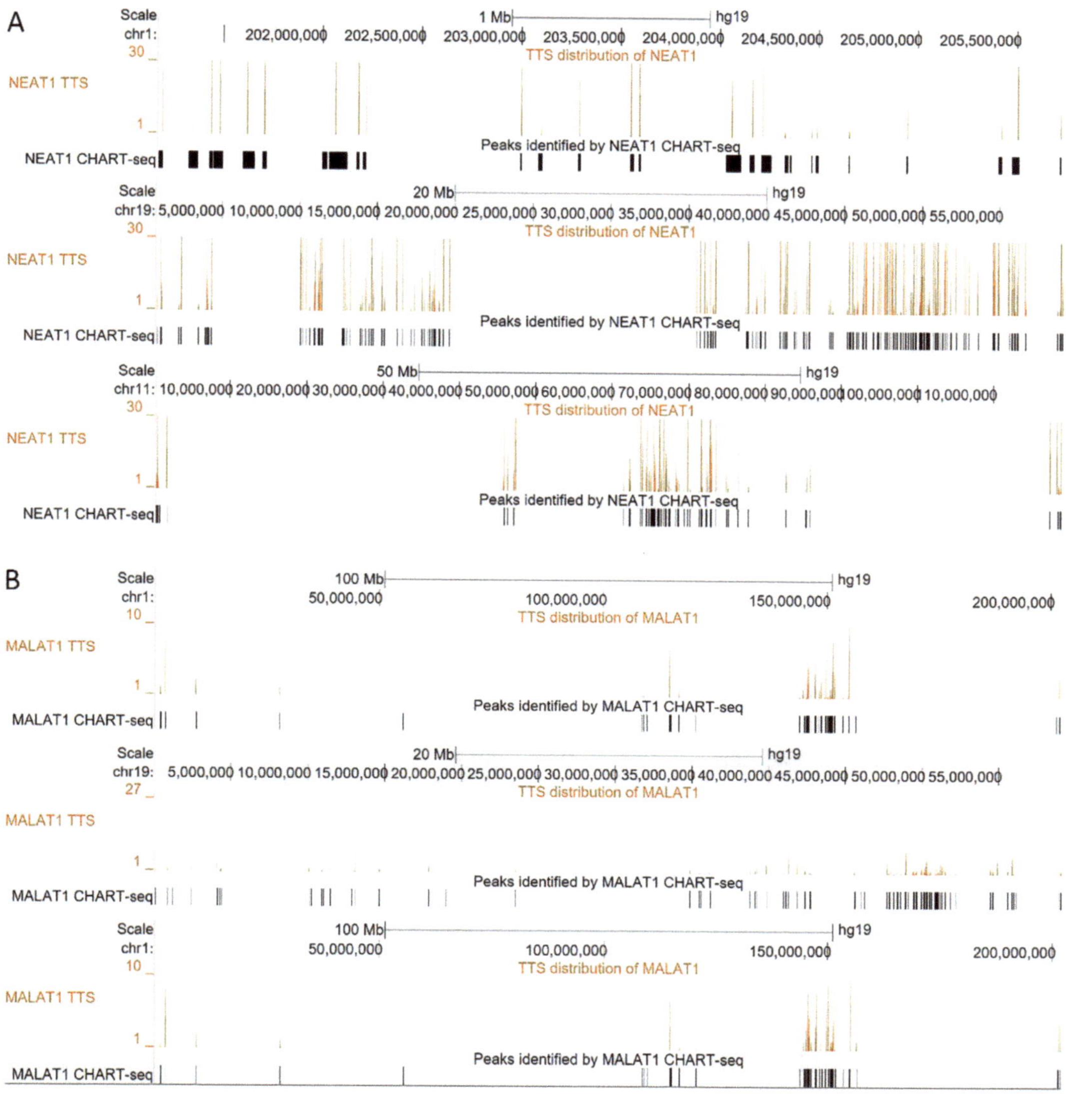

Fig. 15 The DBS distribution of the lncRNA NEAT1 and MALAT1 in two genome regions. The *TFOclass1* files of the two lncRNAs, which contain the genome-wide DBSs in the promoter regions of annotated transcripts, and the experimentally identified (CHART-seq) DNA-binding sites of the two lncRNAs, are uploaded to the UCSC Genome Browser as two custom tracks

LncRNA:DNA-binding prediction can be validated in two ways. First, many imprinted genes and their regulatory lncRNAs have been well known; the user can examine whether the imprinted genes contain DBSs of their regulatory lncRNAs [16]. Second, the user can use CRISPR/Cas9 to knock out the predicted DBD in an lncRNA in some cell lines and perform RNA-seq before and after the knockout to check if exactly the changed gene expression occurs in the predicted target genes (Fig. 17).

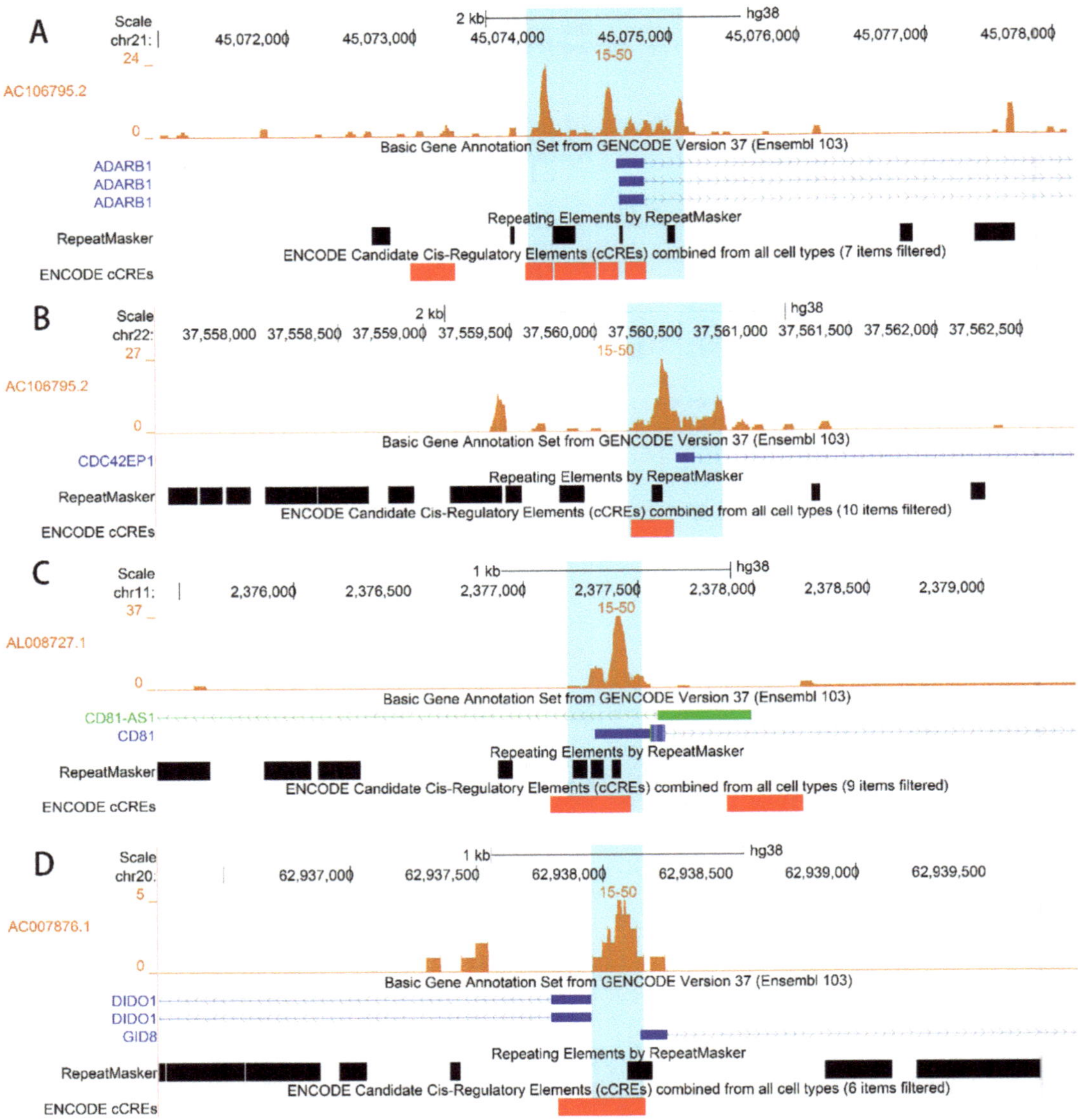

Fig. 16 Examples of co-localization of DBSs, transposable elements, and cis-regulatory elements in the promoter regions of genes. (**a**) The DBSs of AC106795.2 in the promoter region of *ADARB1*. (**b**) The DBSs of AC106795.2 in the promoter region of *CDC42EP1*. (**c**) The DBS of AL008727.1 in the promoter region of *CD81*. (**d**) The DBS of AC007876.1 in the promoter region of *DIDO1* and *GID8*

Multiple methods have been developed to predict DBSs of lncRNAs. Thus, the above three classes of data in the UCSC Genome Browser can also be used to compare DBSs predicted by different methods, by uploading them together with the predicted DBSs as tracks to the UCSC Genome Browser. Note that DBSs to be compared should be predicted with the same length threshold (e.g., Nt=50).

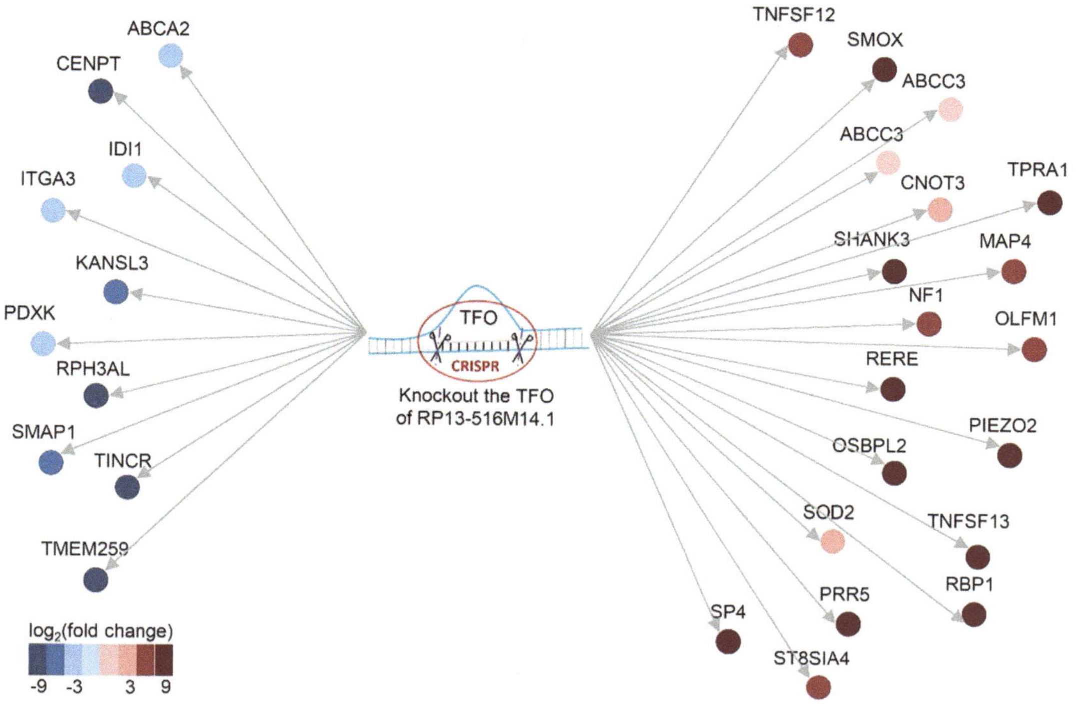

Fig. 17 The changed expression of target genes of the lncRNA RP13-516M14.1 before and after the CRISPR/Cas9 knockout of a short sequence about 100 bp containing the predicted DBD1. Genes whose promoter regions contain DBSs of RP13-516M14.1 are assumed as target genes

5 Notes

Fasim-LongTarget reports fewer triplexes (with less overlapped TTSs) than *LongTarget*, but the number and length of DBSs predicted by the two programs are largely the same [17], guaranteeing that *Fasim-LongTarget* has comparable power to *LongTarget*.

"Nt" and "Offset" are the two most important parameters influencing the number of triplexes and DBSs reported. A larger "Nt" value allows the two programs to report fewer but longer DBSs, while a smaller "Nt" value permits the two programs to report more DBSs, many of which are likely to be short. Different "Nt" values should work with different "Offset" values. Typically, Nt = 30/50/80 work with Offset = 10/15/20. Caution is needed when interpreting short TTSs and DBSs, such as those as short as 30 bp. When comparing DBDs and DBSs predicted by different methods, the threshold controlling DBD length should be the same (e.g., be set to 35 bp or 50 bp). The difference in the threshold affects the comparability of methods.

Since *LongTarget* and *Fasim-LongTarget* use rulesets to examine the binding between every nucleotide, they return both DBDs and DBSs with length and identity (identity is the percentage of paired nucleotides in a triplex). This is also a difference between *LongTarget* and other methods and allows us to define "binding affinity" that indicates the strength of DBSs. According to in vitro studies [12], the strength of a triplex is determined by all paired nucleotides. Thus, it is reasonable to define binding affinity = length * identity. Some researchers may think that a higher identity threshold (e.g., 0.8 or even higher) makes the predicted DBSs more reliable. This is not true. We analyzed the distribution of Kcnq1ot1's DBSs in the large Kcnq1 imprinting region (which contains many known imprinted genes) to explore plausible identity values. We found that a high threshold for identity (e.g., 0.8) will make DBSs in many known imprinted genes fail to be predicted. Upon our analysis of many lncRNAs and in vitro experiments, plausible identity values are most likely between 0.5 and 0.7, thus we set the default threshold at 0.6. A DBS of 100 bp with an identity of 0.6 has a binding affinity of 60, which is a reasonable value for determining target genes. Compared with the length of many TF-DBS, 100 bp is quite long and it is highly unlikely to be generated by chance.

Currently, experimental data such as ENCODE-annotated DNA Methylation signals, Histone Modification signals, and cis-regulatory elements are only available in the hg19 version of the human genome in the UCSC Genome Browser. Although the liftOver function in the UCSC Genome Browser can transfer genomic coordinates from hg38 to hg19, it may be better to use hg19 to make predictions when the primary purpose is to evaluate the reliability of the prediction. On the other hand, using hg38 makes better sense when the primary purpose is to check if DBSs fall in annotated genes or regulatory elements.

Upon the Hoogsteen and reverse Hoogsteen rulesets (currently, there are 24 rulesets), *LongTarget* and *Fasim-LongTarget* often predict multiple DBDs, especially if the input DNA sequence is long. DBD1 (the DBD with more DBSs than any other DBD) is the best DBD, and many other DBDs, especially those with few DBSs, are probably not true. A question is whether DBD2 and DBD3, which have DBSs just fewer than DBD1, are also true ones. Our analyses of KCNQ1OT1 suggest that some long lncRNAs may indeed have multiple DBDs [13], although the examples, evaluation, and validation described in this chapter focus only on DBD1. We conjecture that lncRNAs with multiple DBDs may use different ones to regulate different target genes in different cells.

Finally, we note that although the *LongTarget* website provides an easy way to submit *LongTarget/Fasim-LongTarget* predictions, the computational resources are limited. Thus, tasks may be heavily queued, and the user may have to wait long to obtain the results.

References

1. Derrien T et al (2012) The GENCODE v7 catalog of human long noncoding RNAs: analysis of their gene structure, evolution, and expression. Genome Res 22(9):1775–1789
2. Yue F et al (2014) A comparative encyclopedia of DNA elements in the mouse genome. Nature 515(7527):355–364
3. Breschi A, Gingeras TR, Guigo R (2017) Comparative transcriptomics in human and mouse. Nat Rev Genet 18(7):425–440
4. Hoogsteen K (1963) The crystal and molecular structure of a hydrogen-bonded complex between 1-methylthymine and 9-methyladenine. Acta Crystallographica 16(9):907–916
5. Yap KL et al (2010) Molecular interplay of the noncoding RNA ANRIL and methylated histone H3 lysine 27 by polycomb CBX7 in transcriptional silencing of INK4a. Mol Cell 38(5): 662–674
6. Statello L et al (2021) Gene regulation by long non-coding RNAs and its biological functions. Nat Rev Mol Cell Biol 22(2):96–118
7. Monnier P et al (2013) H19 lncRNA controls gene expression of the imprinted gene network by recruiting MBD1. Proc Natl Acad Sci USA 110(51):20693–20698
8. Flavahan WA, Gaskell E, Bernstein BE (2017) Epigenetic plasticity and the hallmarks of cancer. Science 357(6348)
9. Huarte M (2015) The emerging role of lncRNAs in cancer. Nat Med 21(11): 1253–1261
10. Liu SJ et al (2021) Long noncoding RNAs in cancer metastasis. Nat Rev Cancer 21(7): 446–460
11. Guo W et al (2021) LincRNA-immunity landscape analysis identifies EPIC1 as a regulator of tumor immune evasion and immunotherapy resistance. Sci Adv 7(7)
12. Abu Almakarem AS et al (2012) Comprehensive survey and geometric classification of base triples in RNA structures. Nucleic Acids Res 40(4):1407–1423
13. Liu H, Shang X, Zhu H (2017) LncRNA/DNA binding analysis reveals losses and gains and lineage specificity of genomic imprinting in mammals. Bioinformatics 33(10):1431–1436
14. Buske FA et al (2012) Triplexator: detecting nucleic acid triple helices in genomic and transcriptomic data. Genome Res 22(7): 1372–1381
15. Kuo CC et al (2019) Detection of RNA-DNA binding sites in long noncoding RNAs. Nucleic Acids Res 47(6):e32
16. He S et al (2015) LongTarget: a tool to predict lncRNA DNA-binding motifs and binding sites via Hoogsteen base-pairing analysis. Bioinformatics 31(2):178–186
17. Wen Y et al (2022) Fasim-LongTarget enables fast and accurate genome-wide lncRNA/DNA binding prediction. Comput Struct Biotechnol J 20:3347–3350
18. Warwick T et al (2022) A universal model of RNA.DNA:DNA triplex formation accurately predicts genome-wide RNA-DNA interactions. Brief Bioinform 23(6)
19. Lin J et al (2019) Pipelines for cross-species and genome-wide prediction of long noncoding RNA binding. Nat Protoc 14(3):795–818
20. Nawrocki EP, Kolbe DL, Eddy SR (2009) Infernal 1.0: inference of RNA alignments. Bioinformatics 25(10):1335–1337
21. Nawrocki EP (2014) Annotating functional RNAs in genomes using infernal. Methods Mol Biol 1097:163–197

Chapter 9

Thorough RNA Interactome Data Analysis Based on Direct Duplex Detection Data

Richard A. Schäfer and Björn Voß

Abstract

RNA–RNA interactions play a crucial role in cellular processes by regulating gene expression, RNA stability, and posttranscriptional modifications. These interactions have emerged as key players in the intricate network of RNA-based regulatory mechanisms. In recent years, *Direct Duplex Detection* methods have been widely used to map these interactions on a large scale but require sophisticated data analysis. We describe the analysis of these DDD data using RNAnue to determine RNA–RNA interactions. RNAnue is available at https://github.com/Ibvt/RNAnue.

Key words Direct duplex detection, RNA–RNA interaction, Data analysis

1 Introduction

RNA–RNA interactions play a crucial role in a variety of biological processes. In recent years, *Direct Duplex Detection (DDD) methods* have been shown to provide a holistic approach to detecting RNA–RNA interactions on a global scale. Most commonly, LIGR-seq [1], SPLASH [2], and PARIS [3] are used. However, these methods are inefficient and require sophisticated data analysis [4]. Recently, we introduced RNAnue [5], a comprehensive workflow, to detect RNA–RNA interactions from *Direct Duplex Detection* methods. A typical RNAnue analysis comprises quality and adapter trimming of the sequencing reads, mapping of the reads to the genome, split read calling, clustering, filtering, and statistical analysis. Thereby, individual interactions are found, grouped by location and annotation into transcript interactions, which form the analysis result. In this chapter, we guide the reader through the analysis of DDD data using RNAnue and describe the data formats and pitfalls encountered during the analysis of this data.

Michiaki Hamada et al. (eds.), *RNA-RNA and RNA-DNA Interactions*, Methods in Molecular Biology, vol. 2949,
https://doi.org/10.1007/978-1-0716-4670-0_9,

2 Materials

2.1 Hardware

Linux/Mac operating system with multicore processors. The tasks described were performed on an Intel(R) Xeon(R) CPU E5–2680 v4 @ 2.40 GHz, 128 GB RAM.

2.2 Software

In the following, we assume that the required tools are to be installed into ~/tools/ and added to the PATH environment variable to make them available system-wide.

1. segemehl v0.3.4

```
$ wget https://www.bioinf.uni-leipzig.de/Software/ segemehl/
downloads/segemehl-0.3.4.tar.gz -P ~/tools/ && cd ~/tools
$ tar xvf segemehl-0.3.4.tar.gz && rm segemehl-0.3.4.tar.gz
$ cd segemehl-0.3.4
$ make all
$ export PATH=~/tools/segemehl-0.3.4/:$PATH
```

2. Vienna 2.4.17.

```
$ wget https://www.tbi.univie.ac.at/RNA/download/sourcecode/
2_4_x/ViennaRNA-2.4.17.tar.gz -P ~/tools && cd ~/tools/
$ tar -zxvf ViennaRNA-2.4.17.tar.gz
$ cd ViennaRNA-2.4.17
$ ./configure
$ make
$ sudo make install
$ export PATH=~/tools/ViennaRNA-2.4.7/bin/:$PATH
```

3. Install RNAnue v0.2.5

```
$ wget https://github.com/Ibvt/RNAnue/releases/download/
v0.2.5/
RNAnue-v0.2.5.tar.gz -P ~/tools/ && cd ~/tools/
$ tar xvf RNAnue.tar.gz
$ cd RNAnue/build
$ cmake ../source/ && make
$ export PATH=~/tools/RNAnue/build/:$PATH
```

4. Install Fastq-dump from the NCBI SRA Toolkit. We chose the compiled binaries to download from https://github.com/ncbi/sra-tools/wiki/01.-Downloading-SRA-Toolkit.

```
$ wget http://https://ftp-trace.ncbi.nlm.nih.gov/sra/sdk/
3.0.5/sratoolkit.3.0.5-ubuntu64.tar.gz -P ~/tools/ && cd ~/
tools
$ tar xvf sratoolkit.3.0.5-ubuntu64.tar.gz && rm sratoolk-
it.3.0.5-ubuntu64.tar.gz
$ export PATH=~/tools/sratoolkit.3.0.5-ubuntu64/bin:$PATH
```

3 Methods

3.1 Data Preparation

In this example, we showcase the application of RNAnue using LIGR-seq data which is stored in the GEO datasets (https://www.ncbi.nlm.nih.gov/gds) under accession number: GSE80167. It is to be noted that the input data needs to be stored in folders that contain subfolders for the libraries in which the reads are placed (options –ctrls and –trtms, *see* Subheading 3.3). For example:

ctrls
    library1
    reads.fastq
trtms
    library1
    reads.fastq

1. Download RNA-seq reads

```
$ mkdir -p ~/analysis/data/ctrls/noAMTnoLigase/
$ fastq-dump SRR3361016 && mv SRR3361016.fastq ~/analysis/
data/ctrls/noAMTnoLigase/H293T_001.fastq
$ fastq-dump SRR3361020 && mv SRR3361020.fastq ~/analysis/
data/ctrls/noAMTnoLigase/H293T_002.fastq
$ mkdir -p ~/data/trtms/AMTLigase
$ fastq-dump SRR3361013 && mv SRR3361013.fastq ~/analysis/
data/trtms/AMTLigase/H293T_001.fastq
$ fastq-dump SRR3361017 && mv SRR3361017.fastq ~/analysis/
data/trtms/AMTLigase/H293T_002.fastq
```

2. Download genomic sequences and gene annotation

```
$ mkdir ~/ref/ && cd ~/ref/
$ wget https://ftp.ncbi.nlm.nih.gov/genomes/all/GCF/000/001/
405/
GCF_000001405.40_GRCh38.p14/GCF_000001405.40_GRCh38.p14_geno-
mic.fna.gz
&& gzip -d GCF_000001405.40_GRCh38.p14_genomic.fna.gz
$ wget https://ftp.ncbi.nlm.nih.gov/genomes/all/GCF/000/001/
405/
GCF_000001405.40_GRCh38.p14/GCF_000001405.40_GRCh38.p14_geno-
mic.gff.gz && gzip -d GCF_000001405.40_GRCh38.p14_genomic.gff.
gz
```

3.2 Workflow and Configuration

The workflow of RNAnue consists of the subcalls *preproc*, *align*, *detect*, *clustering*, and *analysis* (**Note 1**) as shown in the following:

(optional) Adapter trimming and quality filtering.

```
$ RNAnue preproc -c params.cfg
```

Aligning sequencing reads to reference genome.

```
$ RNAnue align -c params.cfg
```

Detection of split reads and computation of scores for filtering.

```
$ RNAnue detect -c params.cfg
```

(optional) Clustering of the split reads.

```
$ RNAnue clustering -c params.cfg
```

Analysis of the split reads, annotation, and grouping into transcript interactions.

```
$ RNAnue analysis -c params.cfg
```

Note: All the aforementioned steps can be performed at once.

```
$ RNA complete -c params.cfg
```

RNAnue can be configured through either a configuration file or the command line, where the latter has precedence. Table 1 lists all the available settings.

Table 1
Parameter settings, default values, and descriptions

Parameter	Default	Description
General		
readtype	SE	Read type which can be either single-end (=SE) or paired-end (=PE)
ctrls	–	Absolute path to directory containing the control samples (if available) with additional subdirectories for each library.
trtms	–	Absolute path to directory containing the treatment samples with additional subdirectories for each library
outdir	–	Absolute path to directory to store the results of RNAnue
threads	1	Number of threads to use
quality	20	Lower limit of the average quality of the reads to consider in the analysis (Phred Quality Score)
mapquality	20	Lower limit of the average quality of the alignment
minlen	15	Minimum length of the read to consider for the analysis
splicing	0	Whether (=1) or not (=0) splicing events are considered in the detection of split reads
Preprocessing		
preproc	0	Whether (=1) or not (=0) preprocessing should be performed
modetrm	1	Mode of the read trimming. Only at 5′-end (=0), 3′-end (=1), or both (=2)
adpt5	–	Sequence(s) preceding the 5′-end (N for arbitrary bp) in FASTA format
adpt3	–	Sequence(s) succeeding the 3′-end (N for arbitrary bp) in FASTA format
wtrim	0	Whether (=1) or not (=0) to include window trimming in the preprocessing procedure
mmrate	0.1	Allowed mismatch rate in adapter trimming
wsize	3	Window size if window trimming (see wtrim) is enabled
minovlps	5	Minimum overlaps that are required when merging the paired-end reads
Alignment (segemehl)		
dbref	–	Absolute path to FASTA file of the reference
accuracy	90	Minimum percentage of matches per read in read alignment
minfragsco	15	Minimum score of a spliced fragment
minfraglen	15	Minimum length of spliced fragment
minsplicecov	80	Minimum coverage of spliced transcripts
exclclipping	0	Whether (=0) or not (=1) to exclude soft clipping in results
Split read calling		
sitelenratio	0.3	

(continued)

Table 1
(continued)

Parameter	Default	Description
		Minimum aligned portion of the read when calculating the complementarity (*see* **Note 2**)
cmplmin	0.0	Minimum complementarity (*see* **Note 3**)
nrgmax	0.0	Maximum hybridization energy (*see* **Note 3**)
Clustering		
clust	1	Omit (=0) or include (=1) the clustering procedure
clustdist	0	Minimum distance between the clusters
Analysis		
features	–	Annotations of the organism (*see* **Note 4**)
Output		
stats	1	Produces a statistics files
outcnt	0	Produces a count table in the output
outjgf	0	Produces a JSON graph file of the results

1. Download the predefined configuration file (Fig. 1)

```
$ wget https://raw.githubusercontent.com/Ibvt/RNAnue/master/
build/params_LIGR.cfg -P ~/data/
```

3.3 Output

RNAnue produces outputs on different levels that correspond to the steps in the workflow. In principle, the results are stored in the directory specified in the –outdir parameter. Each step retains the initial folder structure provided by –ctrls and –trtms.

1. The *align* procedure generates the alignment results from segemehl in the folder 'align.' Although segemehl also generates specific BED files, only the SAM files are used in the subsequent analysis. Here, the suffix '_matched' is added to all files.
2. The *detect* step generates two files in SAM format that correspond to the detected split reads. In the '_splits' files, pairs of reads (two lines) correspond to both segments within a split read. Here, the tags `XC` and `XE` are added that correspond to the complementarity and hybridization energy of that split read, respectively. In addition, the tags `XX` and `XY` depict the position of the segment within the read. The '_multsplits' file

```
### BASIC SETTINGS
readtype = SE
ctrls = ~/data/ctrls/noAMTnoLigase/
trtms = ~/data/trtms/AMTLigase/
outdir = ~/data/results/
threads = 30
quality = 20
mapquality = 20
minlen = 20
splicing = 0

### PREPROCESSING
preproc = 0
modetrm = 1
adpt5=
adpt3=
wtrim = 0 wsize = 3
minovlps = 5

### ALIGNMENT (forwarded to segemehl.x)
dbref=~/ref/GCF_000001405.40_GRCh38.p14_genomic.fna
accuracy = 90 alignment
minfragsco = 15
minfraglen = 15
minsplicecov = 80
exclclipping = 0

### SPLIT READ CALLING
sitelenratio = 0.5
cmplmin = 0.0
nrgmax = 0

### CLUSTERING
clust = 1
clustdist = 0

### ANALYSIS
features=~/ref/GCF_000001405.40_GRCh38.p14_genomic.gff

### OUTPUT
stats = 1
outcnt = 1
outjgf = 1
```

Fig. 1 Configuration file used for the analysis of LIGR-seq data

contains split reads that consist of more than two segments (one segment per line).

3. The ‘clustering’ procedure reports a single *clusters.tab* file which is a tab-delimited file of the clustering results. Here,

each line represents a cluster that corresponds to overlapping split reads, defined by the two segments. The columns are defined in the following:

4. The 'analysis' procedure generates '_interactions' files for each library in which each line represents an annotated split read that is mapped to a transcript interaction. The fields are defined as follows:

Field	Description
clustID	Unique identifier of the cluster
fst_seg_chr	Chromosome (accession) of the first strand. This corresponds to the seqid in the provided GFF file
fst_seg_strd	Strand where the first segment is located
fst_seg_strt	Start position of the first segment in the cluster
fst_seg_end	End position of the first segment in the cluster
sec_seg_chr	Chromosome (accession) of the second strand. This corresponds to the seqid in the provided GFF file.
sec_seg_strd	Strand where the second segment is located
sec_seg_strt	Start position of the second segment in the cluster
sec_seg_end	End position of the second segment in the cluster
no_splits	Number of split reads within the cluster
fst_seg_len	Length of the first segment of the cluster
sec_seg_len	Length of the second segment of the cluster

Field	Description
qname	Read/Template identifier
fst_seg_strd	Strand where the first segment is located
fst_seg_strt	Start position of the first segment
fst_seg_end	End position of the first segment
fst_seg_ref	Reference name of the first segment corresponding to the seqid in GFF and/or clusterID
fst_seg_name	Name of the first segment that corresponds to gene name/symbol and/or clusterID
first_seg_bt	Biotype of the annotation transcript (if available)
fst_seg_anno_strd	Strand information of the transcript in the overlapping annotation
fst_seg_prod	Description of the transcript (if available in annotation)

(continued)

Field	Description
fst_seg_ori	Orientation of the segment with respect to annotation (sense/antisense)
sec_seg_strd	Strand where the second segment is located
sec_seg_strt	Start position of the second segment
sec_seg_end	End position of the second segment
sec_seg_ref	Reference name of the second segment corresponding to the seqid in GFF and/or clusterID
sec_seg_name	Name of the second segment that corresponds to gene name/symbol and/or clusterID
sec_seg_bt	Biotype of the annotation transcript (if available)
sec_seg_anno_strd	Strand information of the transcript in the overlapping annotation
sec_seg_prod	Description of the transcript (if available in annotation)
sec_seg_ori	Orientation of the segment with respect to annotation (sense/antisense)
cmpl	Complementarity score of the interaction
fst_seg_compl_aln	Alignment results in the complementarity calculation of the first segment
sec_seg_cmpl_aln	Alignment results in the complementarity calculation of the second segment
mfe	Hybridization energy of the interaction
mfe_struc	Minimum free energy (MFE) structure of interaction in dot-bracket notation

The main result of an RNAnue analysis is transcript interactions. They are stored in the file 'allints.txt' in the same directory. Its entries are structured as described in the following where columns with a prefix <sample> are given for each sample specified in the analysis (within the same file).

Field	Description
fst_rna	Gene/Transcript name of the first interaction partner
sec_rna	Gene/Transcript name of the second interaction partner
fst_rna_ori	Orientation of the interaction partner with respect to annotation (sense/antisense)
sec_rna_ori	Orientation of the interaction partner with respect to annotation (sense/antisense)

(continued)

Field	Description
<sample>_supp_reads	Number of (split)reads that support the interaction
<sample>_ges	Global energy score (gcs) of the interaction
<sample>_ghs	Global hybridization score (ghs) of the interaction
<sample>_gps	Global probability score (gps) of the interaction
<sample>_pval	Statistical significance (p-value) of the interaction
<library>_padj	Benjamini-Hochberg adjusted p-value among the samples

If the option –outcnt is set to 1, RNAnue generates the count table 'counts.txt' in the output directory. It contains the counts of each interaction for each sample and can be used for differential expression analysis. Similarly, −outjgf set to 1 generates a 'graph.json' file that contains the detected interactions in JSON graph format. Finally, −stats set to 1 creates a 'stats.txt' file that contains basic statistics for each step of the analysis.

3.4 RNAnue Using Docker

RNAnue can also be used with Docker. For that, we provide a Docker container at https://hub.docker.com/r/cobirna/rnanue.

1. *Get the Docker container*

```
$ docker pull cobirna/rnanue:0.2.5
```

2. *Create a RNAnue instance and mount the needed volumes*

```
$ docker run -v $PWD/analysis/data:/data -it -d <image_id> /
bin/bash
```

3. *Execute RNAnue*

```
$ docker exec -it <container_id> ./RNAnue <subcall> -c /data/
params_LIGR.cfg
```

4 Notes

1. The clustering procedure of RNAnue allows to refine the interaction landscape and detect novel transcripts, but it also increases the runtime substantially. For that reason, it can be

omitted. Similarly, preprocessing is an optional step and can also be performed beforehand with other tools.

2. The parameter —sitelenratio determines the portion of that read that is included in the complementarity calculation. It should be set to at least 0.3 to prevent high complementarities over short segments.
3. Although the global scores gcs and ghs are robust against outliers, it may be preferable to discard split reads with low complementarity and low hybridization energy. For that, the parameters —cmplmin and —nrgmax can be specified.
4. Although RNAnue is able to select features irrespective of the attributes, it is less error-prone when using common tags (e.g., gene_name, gene_type).

Acknowledgments

German Ministry of Education and Research grants RNAProNet [031L0164A to B.V.] and inteRNAct [031A310 to B.V.].

References

1. Sharma E, Sterne-Weiler T, O'Hanlon D, Blencowe BJ (2016) Global mapping of human RNA-RNA interactions. Mol Cell 62(4): 618–626
2. Aw JGA, Shen Y, Wilm A, Sun M, Lim XN, Boon KL et al (2016) In vivo mapping of eukaryotic RNA interactomes reveals principles of higher-order organization and regulation. Mol Cell 62(4):603–617
3. Lu Z, Zhang QC, Lee B, Flynn RA, Smith MA, Robinson JT et al (2016) RNA duplex map in living cells reveals higher-order transcriptome structure. Cell 165(5):1267–1279
4. Schönberger B, Schaal C, Schäfer R, Voß B (1824) RNA interactomics: recent advances and remaining challenges. F1000Res 2018(7): 1824
5. Schäfer RA, Voß B (2021) RNAnue: efficient data analysis for RNA–RNA interactomics. Nucleic Acids Res 49(10):5493–5501

Chapter 10

Hybrid MPI/OpenMP Implementation of RIblast for Transcriptome-Scale lncRNA–RNA Interaction Prediction

Iñaki Amatria-Barral, Jorge González-Domínguez, and Juan Touriño

Abstract

Because the identification of RNA interaction partners for each long noncoding RNA (lncRNA) has proven to be a powerful approach for inferring lncRNA functions, numerous tools have been developed to predict lncRNA–RNA interactions. However, their application at the transcriptome scale has been hindered by high computational costs. In this chapter, we introduce pRIblast, a hybrid MPI/OpenMP implementation of the RIblast algorithm designed for large-scale lncRNA–RNA interaction analyses. pRIblast leverages cluster environments and supercomputing facilities to reduce the computational costs of these studies dramatically. Moreover, pRIblast has been thoroughly optimized, and it can take on huge datasets that were impossible to analyze with the former RIblast algorithm.

Key words lncRNA, lncRNA–RNA interaction, Transcriptome-scale analysis, High-performance computing, MPI, OpenMP

1 Introduction

Long noncoding RNAs (lncRNAs) are transcripts with no less than 200 nucleotides in length and no protein-coding capacity. Even though lncRNAs were first regarded as transcriptional noise, mounting evidence now suggests that these sequences do have functions and that, indeed, lncRNAs play a key role in many biological processes [1]. For instance, lncRNA regulates gene expression [2], immune response [3], and cell cycle [4].

Although more than 58000 lncRNA genes are encoded by the human genome [5], the vast majority of them are still poorly characterized [6]. And what is worse, the dysfunction of many lncRNAs has been associated with severe diseases, such as cardiac pathologies [7], several types of cancer [8, 9], or Parkinson's [10]. Thus, elucidating lncRNA functions is a hot research topic today.

Michiaki Hamada et al. (eds.), *RNA-RNA and RNA-DNA Interactions*, Methods in Molecular Biology, vol. 2949,
https://doi.org/10.1007/978-1-0716-4670-0_10,

lncRNAs lack sequence and structure conservation [11]. Therefore, sequence similarity search and RNA secondary structure search, while they have achieved success in characterizing the function of protein-coding genes and small RNAs [12, 13], are unsuitable for inferring the functions of lncRNAs. In contrast, because lncRNAs work by being assembled with other proteins or RNAs [14], the identification of interaction partners for each lncRNA has proved to be a successful approach for characterizing the functions of lncRNAs [15, 16].

Several sequencing-based technologies have been developed for the experimental discovery of RNA–RNA interactions. These include RIA-seq [17] and RAP-RNA [18], which are able to exhaustively detect interaction targets of specific lncRNAs. Repeating these experiments across many lncRNAs, however, is very labor intensive. On the other hand, there exist methods, such as PARIS [19] and SPLASH [20], that can comprehensively identify RNA–RNA interactions in vivo. Nonetheless, these techniques have found the most success identifying interactions of ribosomal RNAs and small RNAs. Furthermore, lncRNAs show tissue-specific expression patterns [5, 11], meaning that these experiments should be conducted on various tissues to provide extensive results, which, again, requires hard work.

Since the detection of large-scale lncRNA–RNA interactions through experiments is impracticable, computational prediction has emerged as an indispensable technique. Several tools and procedures have been developed with the sole purpose of predicting RNA–RNA interactions [21, 22], aiming to further progress in the characterization of lncRNA transcripts. For instance, Szcześniak and Makałowska predicted lncRNA–RNA interactions across the human transcriptome using sequence similarity search without consideration of the RNA secondary structure [23]. Nevertheless, benchmarking results have shown that tools that omit the RNA secondary structure have lower discriminatory performance [21, 24]. In other words, the most accurate RNA–RNA interaction prediction results are obtained by tools that consider the RNA secondary structure. Notable examples of such tools include IntaRNA [25], RNAplex [26], RactIP [27], ASSA [28], and RRP [29]. Despite their accuracy, these algorithms require extensive computational resources to find promising lncRNA–RNA interacting pairs due to the consideration of the RNA secondary structure, which significantly impacts execution time [30].

To address the execution time challenge, Fukunaga and Hamada developed RIblast [30], a C++ algorithm that exploits the seed-and-extension heuristic [12]. This approach accelerates the prediction of lncRNA–RNA interacting pairs without compromising prediction accuracy, rivaling the performance of the most accurate tools in the state of the art [24, 30]. However, even with their thorough optimization efforts, RIblast still falls short when it comes to computing lncRNA–RNA interactions at the transcriptome scale. Indeed, Fukunaga and Hamada have already emphasized the need for further acceleration to effectively identify lncRNA–RNA interactions at such a large scale [30]. Therefore, in this chapter, we present pRIblast [31, 32], an enhanced version of RIblast developed to overcome the current execution time limitations. By optimizing the algorithm's runtime and leveraging parallel computing techniques, pRIblast aims to enable the identification of lncRNA–RNA interactions on a broader scale, specifically at the transcriptome level. We demonstrate the improved efficiency and performance of pRIblast and provide practical guidelines for its execution on workstations and computing clusters.

1.1 RIblast

1.1.1 RNA–RNA Interaction Prediction Algorithm

RIblast enumerates potentially interacting segments between a query RNA, q, and a target RNA, t. RIblast uses two energies to determine if two segments, q_s and t_s, interact. These energies are:

- The accessible energy, which indicates segments that are likely to form intermolecular base pairs with other segments. When the accessible energy is low, a segment is more likely to form intermolecular base pairs as it avoids intramolecular base pairing.
- The hybridization energy, which is the free energy released by the intermolecular base pairs between two segments. The hybridization energy is computed as the sum of the stacking energies and loop energies in the intermolecular base pair structure. Intramolecular base pairs are ignored by the hybridization energy algorithm of RIblast.

The interaction energy between q_s and t_s is then computed by adding together the accessible energy of q_s, the accessible energy of t_s, and the hybridization energy between q_s and t_s. RIblast will output segments with particularly low interaction energy as a detected RNA–RNA interaction.

For all-to-all predictions, the calculation time of accessible energies scales linearly with the number of sequences. However, the calculation time of the hybridization energies scales quadratically with the number of sequences. This poses a challenge for comprehensive prediction of lncRNA–RNA interactions. And this is where pRIblast focuses its performance optimization efforts, leveraging parallel computing technologies to reduce the execution time and enable the identification of lncRNA–RNA interactions at the transcriptome scale.

1.1.2 RIblast Overview

RIblast functions in two steps: a database construction step and an RNA interaction search step.

In the database construction step, RIblast precalculates the accessible energies and the results of searching for short substrings of a target RNA dataset. This is done by first computing approximate accessible energies of the RNA sequences and then encoding the sequences, building a suffix array, and constructing a lookup table of search results. All these data (i.e., the accessible energies, the encoded sequences, the suffix array, and the lookup table) are stored in a database that will later be used in the RNA interaction search step. Building a database of RNA sequences allows RIblast to:

- Accelerate the subsequent RNA interaction search step. The accessibility and lookup table computations do not need to be repeated for every query sequence, resulting in substantial computational time savings.
- Enable the efficient reuse of data. Once a database is built, the RNA interaction search can be performed with different query sequences on the same target dataset without having to recompute accessibility and lookup table information. This proves particularly useful when analyzing large datasets, such as the complete human RNA transcriptome.

In the RNA interaction search step, RIblast finds interactions between the target RNA sequences in a database and a given set of query sequences. Here, RIblast first computes approximate accessible energies and constructs a suffix array for every sequence in the query RNA dataset. Second, RIblast finds seed regions with a hybridization energy that does not exceed an energy threshold, T_1. This is done based on the suffix array of the queries and the target database. Third, RIblast computes the interaction energy of the seeds and removes those with energies that exceed 0 kcal/mol. Fourth, seeds are extended to find interactions with no gaps. Fifth, overlapping interactions are removed, as well as interactions with energies exceeding a threshold, T_2. And sixth, interactions are extended from seed regions with a gap and stored in a comma-separated output text file.

For further information on the functioning and design of the RIblast algorithm, we refer the interested readers to the original paper of RIblast [30], especially those seeking to optimize the search parameters of the algorithm for their specific use case.

1.2 pRIblast

pRIblast is a hybrid MPI/OpenMP implementation of the RIblast algorithm specifically designed to enable large-scale lncRNA–RNA interaction analyses. While pRIblast shares the same prediction algorithm as RIblast, it leverages the power of high-performance computing to address the increasing demand for comprehensive lncRNA–RNA interaction analysis.

pRIblast also functions in two steps: a database construction step and an RNA interaction search step. These steps have undergone extensive optimization efforts to maximize performance, including static analysis improvements, data storage and memory management optimizations, and the utilization of advanced parallelization techniques. These optimizations collectively enhance the efficiency and scalability of pRIblast, enabling it to handle huge lncRNA–RNA datasets.

By incorporating pRIblast into their research workflow, scientists can benefit from its ability to dramatically reduce the computational costs associated with large-scale lncRNA–RNA interaction analysis. This empowers researchers to explore the functional implications of lncRNAs at the transcriptome scale and uncover novel insights into their roles in cellular processes.

1.2.1 MPI, OpenMP, and pRIblast's Target Architecture

pRIblast is designed to target distributed-memory architectures, such as computing clusters or supercomputers. These architectures consist of multiple computing nodes, with each node comprising a complete individual computer, connected through a high-performance interconnection network.

In this distributed-memory scenario, pRIblast divides the workload among multiple MPI (Message-Passing Interface) processes that run on the different nodes of the cluster. MPI [33] is the de facto standard for developing parallel programs that run on distributed-memory systems. It provides a portable, highly optimized, and flexible message-passing protocol, facilitating efficient communication and coordination between processes.

Once the workload is distributed among the nodes of the cluster using MPI, each process spawns OpenMP threads to exploit all the CPU cores within each node and accelerate the computation, ideally by a factor of $N \times C$ (where N is the number of nodes and C is the number of cores per node). OpenMP [34] is a collection of compiler directives and functions that offers a scalable, up-to-date, and portable application programming interface for shared-memory systems programming.

For example, in a computing cluster consisting of two nodes, each with four CPU cores, MPI would be used to coordinate two processes running on the two nodes. Each process would then spawn four OpenMP threads to fully exploit the eight cores within this two-node setup, theoretically accelerating a given computation by a factor of 8. In the context of RNA–RNA interaction prediction, pRIblast uses MPI to distribute RNA sequences among the nodes of the cluster, and OpenMP is employed to process these sequences in parallel across all the CPU cores of the cluster.

This paradigm is flexible and allows pRIblast to not only exploit distributed-memory architectures but also standard desktop computers and workstations. Therefore, pRIblast can accelerate the prediction of lncRNA–RNA interacting pairs even on a single machine with multiple CPU cores, providing a more modest configuration for users who do not have access to large-scale computing resources. This demonstrates the versatility of pRIblast, making it suitable for a wide range of computing environments and accommodating users with varying resources and requirements.

1.2.2 Database Paging Mechanism

pRIblast introduces a highly effective optimization known as the database paging mechanism to overcome the memory limitations of the former RIblast algorithm. In RIblast, results are written to the output file only after comparing all target sequences against a specific RNA query sequence. This approach can lead to memory overflow if the target database is large or if a query segment produces a significant number of prediction candidates.

To address this limitation, pRIblast divides the target database into smaller, independent subsets called "pages." These pages are designed to fit within the available memory space of a single compute node and prevent memory overflow. Instead of comparing query sequences against the entire target database in a single round, pRIblast makes predictions against subsets of the database represented by these pages.

Implementing the database paging mechanism required slight modifications to the database construction step described before. Specifically, the database construction step was adjusted to read chunks of a fixed size that correspond to the pages used in the RNA interaction search step. Nevertheless, databases constructed without the modified version of the construction step remain fully compatible with the new pRIblast workflow.

The database paging mechanism in pRIblast significantly improves the efficiency, scalability, and memory management of the algorithm. By dividing the target database into manageable subsets, pRIblast minimizes memory pressure and prevents overflow, enabling researchers to analyze large-scale lncRNA–RNA interactions without compromising computational resources.

1.2.3 pRIblast in Figures

To demonstrate the enhanced performance of pRIblast compared to RIblast, we present a comparison of execution times for predicting lncRNA–RNA interactions on five noncoding RNA transcriptomes. These experiments were conducted using a computing cluster comprising 16 nodes, each equipped with 16 CPU cores. Table 1 provides an overview of the execution times obtained from these experiments, showcasing the efficiency gains achieved by pRIblast.

The table shows that pRIblast consistently outperforms RIblast across all five transcriptomes. Leveraging an optimized algorithm and parallel computing techniques, pRIblast enables significantly faster database construction, completing the task up to 318 times faster than RIblast. Similarly, pRIblast accelerates the prediction of RNA–RNA interactions by up to 135 times compared to RIblast. These advancements capitalize on the full computational power of pRIblast, facilitating rapid and comprehensive analysis of lncRNA–RNA interactions.

Table 1
Execution time comparison between RIblast and pRIblast on five non-coding RNA transcriptomes and a computing cluster with 16 nodes, each equipped with 16 CPU cores

	RIblast		pRIblast	
Transcriptome	DB const.	RNA–RNA search	DB const.	RNA–RNA search
Lepidothrix C.	8 m 0 s	2 h 6 m 42 s	6 s	3 m 37 s
Ursus A.	13 m 56 s	4 h 38 m 34 s	5 s	1 m 59 s
Drosophila M.	41 m 36 s	14 h 43 m 49 s	30 s	9 m 55 s
Anser B.	2 h 8 m 13 s	22d 8 h 13 m 46 s*	24 s	3 h 56 m 6 s
Anas P. P.	3 h 14 m 41 s	101 d 10 h 52 m 55 s*	36 s	20 h 7 m 45 s

* Estimate of the runtime, RIblast cannot compute these datasets because it runs out of memory

Furthermore, the database paging mechanism employed in pRIblast proves to be a crucial optimization, as highlighted in Table 1. This mechanism enables pRIblast to successfully complete two transcriptome-scale analyses that would have been infeasible with RIblast due to memory limitations.

For further information on the technical details behind pRIblast, we refer the interested readers to the original papers of pRIblast [31, 32].

2 Materials

2.1 Hardware Requirements

Choose one of the following:

- An x86/x86_64, SSE2-equipped processor, i.e., virtually any modern x86 processor (year 2000 onward)
- A cluster of x86/x86_64, SSE2-equipped processors, i.e., virtually any computing cluster

2.2 Software Requirements

1. A Linux-/Unix-like operating system, e.g., Ubuntu 20.04
2. Wget, e.g., GNU Wget v1.20.3
3. Gzip, e.g., GNU Gzip v1.10
4. Git, e.g., Git v2.25.1
5. Make, e.g., GNU Make v4.2.1
6. A C++17 compliant and OpenMP-enabled C++ compiler, e.g., GCC v9.4.0
7. An MPI-3 compliant MPI library, e.g., OpenMPI v4.0.3
8. Python 3, e.g., Python v3.8.10

For users executing pRIblast on an Ubuntu desktop/workstation, all these dependencies may be installed at once using the following command:

```
$ sudo apt update && \
 sudo apt install -y wget gzip git build-essential
openmpi-bin libopenmpi-dev python3
```

Users executing pRIblast on a computing cluster will likely have all these packages installed by default. However, they will need to load them explicitly using the appropriate module management tool in place at their system. In the computing cluster used to demonstrate the methods in this chapter, this software is Environment Modules[1] (as it is in most computing clusters worldwide), and we load the required software by executing the following command:

```
$ module load gcc/system openmpi/4.1.4 python/
3.9.9
```

Please note that different installations of Environment Modules may have different module names. The provided command is specific to our system. If you are unsure about how to load the required software, please consult your system administrator.

3 Methods

This section will show how to:

- Install pRIblast on your system.
- Download a noncoding RNA transcriptome and prepare it for analysis.
- Run pRIblast on a:
  - Desktop computer or workstation.
  - Computing cluster.

3.1 Installing pRIblast

This section provides step-by-step instructions for installing pRIblast on your system. It ensures that users have pRIblast installed and ready to use for the subsequent steps.

1. Navigate to your `HOME` directory and create an `opt` folder where you will store your local installation of pRIblast.

```
$ cd ~
$ mkdir -p opt
```

[1] https://lmod.readthedocs.io/en/latest/

2. Navigate to the newly created folder and clone pRIblast into your local machine.

```
$ cd opt
$ git clone https://github.com/UDC-GAC/pRIblast
```

3. Navigate to the root folder of pRIblast and compile the application (*see* **Note 1** below).

```
$ cd pRIblast
$ make -j$(getconf _NPROCESSORS_ONLN)
```

4. Update your PATH variable so that your system can find the pRIblast executable file automatically. We recommend changing the name of the executable file, too.

```
$ mv target/pRIblast.release target/pRIblast
$ echo "export PATH=\$PATH:$(pwd)/target" >> ~/.
bashrc
```

5. Restart your shell's session, and you will be able to invoke pRIblast from anywhere in your system.

```
$ cd ~
$ source .bashrc
$ pRIblast
```

3.2 Data Preparation

In this section, the focus is on preparing the data for analysis with pRIblast. It guides users through creating a data directory, downloading a specific noncoding RNA transcriptome, decompressing the downloaded file, and splitting the transcriptome into separate files for RNA and lncRNA sequences.

1. Navigate to your HOME directory and create a data folder where you will store the transcriptome executed in the following sections of the chapter.

```
$ cd ~
$ mkdir -p data/pRIblast/input
```

2. Navigate to the newly created folder and download the *Lepidothrix coronata* noncoding RNA transcriptome from the Ensembl genome browser 109 [35].

```
$ cd data/pRIblast/input
$ wget http://ftp.ensembl.org/pub/release-109/
fasta/lepidot\
        hrix_coronata/ncrna/Lepidothrix_coronata.
        Lepidothrix\
         _coronata-1.0.ncrna.fa.gz -O lepidothrix.
         fa.gz
```

3. Decompress the transcriptome.

```
$ gzip -d lepidothrix.fa.gz
```

4. Run the script provided alongside pRIblast to split `lepidothrix.fa` into two separate files: `lepidothrix_RNA.fa`, which will hold all the RNA sequences in the transcriptome and will be used later to construct a database, and `lepidothrix_lncRNA.fa`, which will hold only the lncRNA sequences in the transcriptome and will be used later to predict interactions against `lepidothrix_RNA.fa`.

```
$ python3 ~/opt/pRIblast/scripts/split_tran-
scriptome.py lepidothrix.fa
```

3.3 Running pRIblast on a Desktop Computer or Workstation

This section provides instructions for running pRIblast on a desktop computer or workstation.

1. Navigate to your `HOME` directory and create an `output` folder inside the `data` folder where you will store the results of pRIblast.

```
$ cd ~
$ mkdir -p data/pRIblast/output
```

2. Create a temporary directory inside `/tmp` for pRIblast to store intermediate result files.

```
$ mkdir -p /tmp/pRIblast
```

3. Build a database of RNA sequences, `lepidothrix-db`, with the complete RNA transcriptome downloaded in the previous section.

```
$ pRIblast db -i data/pRIblast/input/lepido-
thrix_RNA.fa \
        -o data/pRIblast/output/lepidothrix-db \
        -p /tmp/pRIblast
```

4. Predict interactions between the lncRNA sequences in the *Lepidothrix coronata* transcriptome and the RNA sequences in the newly created database (*see* **Note 2** below). Predicted interactions will be stored in the file `lepidothrix-pred.txt`.

```
$ pRIblast ris -i data/pRIblast/input/lepido-
thrix_lncRNA.fa \
       -o data/pRIblast/output/lepidothrix-pred.
txt \
       -d data/pRIblast/output/lepidothrix-db \
       -p /tmp/pRIblast
```

The previous commands will launch one pRIblast process and spawn as many threads as there are CPU cores available in your computer (*see* **Note 3** below). This demonstrates the flexibility provided by pRIblast and how it seamlessly scales from standard desktop computers to supercomputers.

3.4 Running pRIblast on a Computing Cluster

This section provides instructions for running pRIblast on a computing cluster. But please note that the method described here is specifically designed for clusters managed with SLURM[2] (about 60% of the supercomputers in the TOP500 list) and Environment Modules. However, it can serve as a solid foundation for creating execution scripts in clusters managed with other clustering software.

1. Navigate to your `HOME` directory and create an `output` folder inside the `data` folder where you will store the results of pRIblast.

```
$ cd ~
$ mkdir -p data/pRIblast/output
```

2. Create a `scripts` directory where you will store the batch script used to submit pRIblast for execution on the cluster.

```
$ mkdir -p scripts/pRIblast
```

3. Navigate to the `scripts` folder and create the following batch script with the name `lepidothrix.sbatch`. This script will first load all the required software to run pRIblast. Then, it will create a database of RNA sequences, `lepidothrix-db`, using all the RNA sequences in the *Lepidothrix coronata* transcriptome. And finally, it will predict lncRNA–RNA interactions between the RNA sequences in the database and the lncRNA sequences in the transcriptome.

```
$ cd scripts/pRIblast
$ cat <<EOF > lepidothrix.sbatch
#!/bin/bash
#SBATCH --nodes=16
#SBATCH --ntasks-per-node=1
#SBATCH --cpus-per-task=16
#SBATCH --mem-per-cpu=4G
#SBATCH --time=00:20:00

module load gcc/system openmpi/4.1.4

export OMP_NUM_THREADS=\$SLURM_CPUS_PER_TASK
srun pRIblast db -i ~/data/pRIblast/input/lepido-
thrix_RNA.fa \
      -o ~/data/pRIblast/output/lepidothrix-db \
      -p \$TMPDIR
```

[2] https://slurm.schedmd.com/documentation.html

```
srun pRIblast ris -i ~/data/pRIblast/input/lepi-
dothrix_lncRNA.fa \
    -o ~/data/pRIblast/output/lepidothrix-pred.
txt \
    -d ~/data/pRIblast/output/lepidothrix-db \
    -p \$TMPDIR
EOF
```

4. Use `sbatch` to submit the script for execution on the cluster (*see* **Notes 2** and **4** below).

```
$ sbatch lepidothrix.sbatch
```

This last command initiates the execution of `lepidothrix.sbatch` on the cluster in batch mode (i.e., noninteractive, offline execution). The script launches pRIblast with 16 processes, each spawning 16 threads, distributed across 16 different nodes of the cluster, thereby using a total of 256 CPU cores. Additionally, the script allocates 4 GB of main memory per thread and sets a time limit of 20 minutes for the execution. To optimize performance and reduce input and output latencies, the `$TMPDIR` environment variable points to a local temporary directory on each node's local file system.

4 Notes

The following notes provide additional information and troubleshooting tips for potential issues that users may encounter during the installation and execution of pRIblast.

1. If the compilation fails and throws any of the following error messages:
   - `make: command not found`, or
   - `make: mpic++: command not found`,

   make sure you have installed all the required software listed in Subheading 2.2 before compiling pRIblast. Users compiling pRIblast in a cluster make sure you have loaded all the required software using the appropriate module management tool.
2. If the interaction prediction step fails with an error message of the form:
   - `segmentation fault (core dumped)`,
   - `fatal error: out of memory`, or
   - `error: detected oom-kill event(s)`,

   this means that pRIblast ran out of memory while finding RNA interactions. Rebuild the target database, but this time set a page size (i.e., append the option `-c INT`) to reduce the maximum number of sequences that are processed simultaneously by the algorithm. We have found that a page size of 250 sequences reduces the memory consumption of the algorithm below 4 GB per thread. If you still encounter out-of-memory errors, reconstruct the database with a smaller page size (i.e., `-c 150`).

3. To control the number of threads used by pRIblast on a desktop computer, users may set the environment variable `OMP_NUM_THREADS` accordingly. For instance, to limit the number of threads spawned by pRIblast to 4, execute the application as follows:

```
$ OMP_NUM_THREADS=4 pRIblast ...
```

4. If the submission fails with an error message of the form `sbatch: error: batch job submission failed`, please check that the resources requested by the batch script do not violate any hard limit or accounting/quality of service policy of your system (e.g., job submission limits or time limits). For instance, it may happen that your cluster does not allow configurations where memory per CPU core exceeds a threshold of 2 GB. If you are unsure, please contact your system administrator for further assistance.

Acknowledgements

We sincerely thank Dr. Michiaki Hamada, Dr. Tsukasa Fukunaga, and Dr. Chao Zeng for inviting us to write this chapter for Methods in Molecular Biology. This work was supported by the Ministry of Science and Innovation of Spain (PID2019-104184RB-I00/AEI/10.13039/501100011033), by the Ministry of Education of Spain (FPU21/00491), and by Xunta de Galicia (Consolidation Program of Competitive Reference Groups, ref. ED431C 2021/30).

References

1. Zhu JJ, Fu HJ, Wu YG, Zheng XF (2013) Function of lncRNAs and approaches to lncRNA-protein interactions. Sci China Life Sci 56:876–885
2. Fatica A, Bozzoni I (2014) Long non-coding RNAs: new players in cell differentiation and development. Nat Rev Genet 15(1):7–21
3. Ouyang J, Hu J, Chen J-L (2016) lncRNAs regulate the innate immune response to viral infection. Wiley Interdiscip Rev RNA 7(1): 129–143
4. Kitagawa M, Kitagawa K, Kotake Y, Niida H, Ohhata T (2013) Cell cycle regulation by long non-coding RNAs. Cell Mol Life Sci 70:4785–4794
5. Iyer MK, Niknafs YS, Malik R, Singhal U, Sahu A, Hosono Y, Barrette TR, Prensner JR, Evans JR Zhao S et al (2015) The landscape of long noncoding RNAs in the human transcriptome. Nat Genet 47(3):199–208
6. de Hoon M, Shin JW, Carninci P (2015) Paradigm shifts in genomics through the FANTOM projects. Mammalian Genome 26(9–10):391–402

7. Hobuß L, Bär C, Thum T (2019) Long non-coding RNAs: at the heart of cardiac dysfunction? Front Physiol 10:30
8. Tornesello ML, Faraonio R, Buonaguro L, Annunziata C, Starita N, Cerasuolo A, Pezzuto F, Tornesello AL, Buonaguro FM (2020) The role of microRNAs, long non-coding RNAs, and circular RNAs in cervical cancer. Front Oncol 10:150
9. Dong X, He X, Guan A, Huang W, Jia H, Huang Y, Chen S, Zhang Z, Gao J, Wang H (2019) Long non-coding RNA Hotair promotes gastric cancer progression via miR-217-GPC5 axis. Life Sci 217:271–282
10. Elkouris M, Kouroupi G, Vourvoukelis A, Papagiannakis N, Kaltezioti V, Matsas R, Stefanis L, Xilouri M, Politis PK (2019) Long non-coding RNAs associated with neurodegeneration-linked genes are reduced in Parkinson's disease patients. Front Cell Neurosci 13:58
11. Cabili MN, Trapnell C, Goff L, Koziol M, Tazon-Vega B, Regev A, Rinn JL (2011) Integrative annotation of human large intergenic noncoding RNAs reveals global properties and specific subclasses. Genes Dev 25(18): 1915–1927
12. Altschul SF, Gish W, Miller W, Myers EW, Lipman DJ (1990) Basic local alignment search tool. J Mol Biol 215(3):403–410
13. Nawrocki EP, Eddy SR (2013) Infernal 1.1: 100-fold faster RNA homology searches. Bioinformatics 29(22):2933–2935
14. Hirose T, Mishima Y, Tomari Y (2014) Elements and machinery of non-coding RNAs: toward their taxonomy. EMBO Rep 15(5): 489–507
15. Abdelmohsen K, Panda AC, Kang M-J, Guo R, Kim J, Grammatikakis I, Yoon J-H, Dudekula DB, Noh JH, Yang X et al (2014) 7SL RNA represses p53 translation by competing with HuR. Nucleic Acids Res 42(15):10099–10111
16. Gong C, Maquat LE (2011) lncRNAs transactivate STAU1-mediated mRNA decay by duplexing with 3′ UTRs via Alu elements. Nature 470(7333):284–288
17. Kretz M, Siprashvili Z, Chu C, Webster DE, Zehnder A, Qu K, Lee CS, Flockhart RJ, Groff AF, Chow J et al (2013) Control of somatic tissue differentiation by the long non-coding RNA TINCR. Nature 493(7431):231–235
18. Engreitz JM, Sirokman K, McDonel P, Shishkin AA, Surka C, Russell P, Grossman SR, Chow AY, Guttman M, Lander ES (2014) RNA-RNA interactions enable specific targeting of noncoding RNAs to nascent Pre-mRNAs and chromatin sites. Cell 159(1): 188–199
19. Lu Z, Zhang QC, Lee B, Flynn RA, Smith MA, Robinson JT, Davidovich C, Gooding AR, Goodrich KJ, Mattick JS et al (2016) RNA duplex map in living cells reveals higher-order transcriptome structure. Cell 165(5): 1267–1279
20. Aw JGA, Shen Y, Wilm A, Sun M, Lim XN, Boon K-L, Tapsin S, Chan Y-S, Tan C-P, Sim AY et al (2016) In vivo mapping of eukaryotic RNA interactomes reveals principles of higher-order organization and regulation. Mol Cell 62(4):603–617
21. Lai D, Meyer IM (2016) A comprehensive comparison of general RNA–RNA interaction prediction methods. Nucleic Acids Res 44(7): e61
22. Umu SU, Gardner PP (2017) A comprehensive benchmark of RNA–RNA interaction prediction tools for all domains of life. Bioinformatics 33(7):988–996
23. Szcześniak MW, Makałowska I (2016) lncRNA–RNA interactions across the human transcriptome. PLoS One 11(3):e0150353
24. Antonov IV, Mazurov E, Borodovsky M, Medvedeva YA (2019) Prediction of lncRNAs and their interactions with nucleic acids: benchmarking bioinformatics tools. Briefings Bioinform 20(2):551–564
25. Busch A, Richter AS, Backofen R (2008) IntaRNA: efficient prediction of bacterial sRNA targets incorporating target site accessibility and seed regions. Bioinformatics 24(24): 2849–2856
26. Tafer H, Hofacker IL (2008) RNAplex: a fast tool for RNA–RNA interaction search. Bioinformatics 24(22):2657–2663
27. Kato Y, Sato K, Hamada M, Watanabe Y, Asai K, Akutsu T (2010) RactIP: fast and accurate prediction of RNA-RNA interaction using integer programming. Bioinformatics 26(18): i460–i466
28. Antonov I, Marakhonov A, Zamkova M, Medvedeva Y (2018) ASSA: fast identification of statistically significant interactions between long RNAs. J Bioinform Comput Biol 16(1): 1840001
29. Terai G, Iwakiri J, Kameda T, Hamada M, Asai K (2016) Comprehensive prediction of lncRNA–RNA interactions in human transcriptome. BMC Genom 17:153–164
30. Fukunaga T, Hamada M (2017) RIblast: an ultrafast RNA–RNA interaction prediction system based on a seed-and-extension approach. Bioinformatics 33(17):2666–2674

31. Amatria-Barral I, González-Domínguez J, Touriño J (2023) pRIblast: a highly efficient parallel application for comprehensive lncRNA–RNA interaction prediction. Fut Gener Comput Syst 138:270–279
32. Amatria-Barral I, González-Domínguez J, Touriño J (2023) Parallel construction of RNA databases for extensive lncRNA–RNA interaction prediction. In: Proceedings of the 38th ACM/SIGAPP symposium on applied computing, SAC '23 (Tallinn, Estonia), pp 555–558
33. The MPI Forum (2012) MPI: A Message-Passing Interface Standard (v3.0). https://www.mpi-forum.org/docs/mpi-3.0/mpi30-report.pdf (visited on 22/06/2023)
34. OpenMP Architecture Review Board (2015) OpenMP Application Programming Interface (v4.5). https://www.openmp.org/wp-content/uploads/openmp-4.5.pdf (visited on 22/06/2023)
35. Martin FJ, Amode MR, Aneja A, Austine-Orimoloye O, Azov AG, Barnes I, Becker A, Bennett R, Berry A, Bhai J et al (2023) Ensembl 2023. Nucleic Acids Res 51(D1): D933–D941

Chapter 11

Target RNA Recognition of MicroRNA-Mediated RNA Silencing

Tomoko Takahashi

Abstract

Gene expression undergoes precise regulation across various stages, from transcription through translation to degradation. An integral player in this regulation is microRNA (miRNA), a small noncoding RNA that modulates gene expression by inducing mRNA degradation or translational repression through sequence complementarities. This section describes experimental methods and useful web tools for predicting and validating RNA–RNA interactions in the context of miRNA-mediated RNA silencing. These methods are also applicable for analyzing potential off-target effects induced by small interfering RNA (siRNA).

Key words RNA silencing, MicroRNA, siRNA, Target RNA, Seed region, Sequence complementarity, Prediction and validation

1 Introduction

RNA silencing is a posttranscriptional gene silencing mechanism directed by small noncoding RNAs (ncRNAs) such as microRNAs (miRNAs) or small interfering RNAs (siRNAs) [1–3]. miRNAs, approximately 22 nucleotides in length, are encoded in the genomes of diverse organisms. According to the miRNA database, miRBase, the human genome encodes 1917 precursors [4]. miRNAs are transcribed from the genome as primary transcripts (pri-miRNAs) by RNA polymerase II and undergo processing in the nucleus by the endoribonuclease Drosha, resulting in miRNA precursors (pre-miRNAs) [5–7]. These pre-miRNAs are then exported to the cytoplasm by exportin-5 and further processed into miRNA duplexes by the endoribonuclease Dicer [8, 9]. Within the RNA-induced silencing complex (RISC), a key player is Argonaute (AGO), with the mature miRNA forming a base pair with the target RNA at the seed region (positions 2–7 or 2–8 from the 5′ end of mature miRNAs) [10–12]. This RNA–RNA interaction

Michiaki Hamada et al. (eds.), *RNA-RNA and RNA-DNA Interactions*, Methods in Molecular Biology, vol. 2949,
https://doi.org/10.1007/978-1-0716-4670-0_11,

induces RNA degradation or translational repression, orchestrated by factors recruited to RISC [13].

The human genome encodes four AGOs (AGO1–4), with AGO2 cleaving target RNAs through a perfect base pair with the entire sequence, a mechanism termed RNA interference (RNAi) [3]. This mechanism is widely employed in laboratories for targeted gene knockdown through the introduction of chemically synthesized siRNA into cells. Beyond laboratory applications, siRNA has found clinical utility. In 2018, Patisiran, the first siRNA medicine, gained FDA approval for treating hereditary transthyretin-mediated amyloidosis [14]. Well-designed and chemically synthesized siRNAs are envisioned to selectively target genes of interest with high specificity upon cell introduction. In contrast, endogenous miRNAs exhibit a broader impact, simultaneously targeting hundreds of genes [15, 16]. Although the 3′ untranslated region (3′UTR) is the most characterized region targeted by miRNAs [15, 16], evidence indicates that miRNA target sites within coding sequences (CDS) and introns, exerting functional repression [17]. The interaction between miRNAs and their target RNAs significantly influences cellular functions.

This section describes experimental methods and valuable web tools for predicting and validating RNA–RNA interactions pivotal to miRNA-mediated RNA silencing. Additionally, these methods are applicable for assessing potential off-target effects induced by siRNA.

2 Materials

2.1 Cell Culture

1. Human NTERA2/D1 cells (obtained from ATCC) or other adherent culture cells.
2. 12- or 24-well plates for cell culture.
3. Dulbecco's Modified Eagle's Medium (DMEM).
4. Fetal bovine serum (FBS).
5. Penicillin–streptomycin solution (×100) (10,000 U/mL of penicillin and 10,000 μg/mL of streptomycin).
6. CO_2 incubator.
7. Liquefied carbon dioxide.

2.2 RNA Purification

1. FastGene RNA Premium Kit.
2. FastGene miRNA enhancer.
3. 2-Mercaptoethanol.
4. NanoDrop One (Thermo Fisher Scientific).

2.3 Quantitative Reverse Transcription (qRT)-PCR

1. High-Capacity cDNA Reverse Transcription Kit (Thermo Fisher Scientific).
2. Stem loop RT primer for mature miR-106b: gtcgtatccagtgcagggtccgaggtattcgcactggatacgacATCTGC.
3. Stem loop RT primer for mature miR-19b: gtcgtatccagtgcagggtccgaggtattcgcactggatacgacTCAGTT.
4. T100 Thermal Cycler (BioRad).
5. KAPA SYBR Fast qPCR kit (KAPA Biosystems).
6. miR-106b qRT-PCR forward primer: gccgcTAAAGTGCTGACAGT.
7. miR-19b qRT-PCR forward primer: gccgcTGTGCAAATCCATGCAA.
8. Universal reverse primer: gtgcagggtccgaggt.
9. β-Tubulin forward primer: CTGGCACCATGGACTCTG.
10. β-Tubulin reverse primer: TCGGCTCCCTCTGTGTAG.
11. QuantStudio 3 Real-Time PCR system (Thermo Fisher Scientific).

2.4 Plasmid Transfection and Dual Luciferase Reporter Assay

1. OPTI-MEM (Thermo Fisher Scientific).
2. Polyethylenimine (PEI) Max or Lipofectamine (Thermo Fisher Scientific).
3. Passive lysis buffer (Promega).
4. Dual-luciferase reporter assay system (Promega).

3 Methods

This section describes the method for evaluating endogenous miRNA expression levels following viral infection using qRT-PCR to facilitate the investigation of miRNA-target RNA interaction [18–21] (*see* **Note 1**). Assessing the endogenous expression levels or expression patterns of miRNAs in response to stimuli in the cells of interest enables the selection of an optimal experimental method for studying miRNA-target RNA interactions and their effects (*see* **Note 2**).

3.1 Measurement of Endogenous Mature miRNA Expression Levels by qRT-PCR

1. Culture cells in Dulbecco's Modified Eagle's Medium (DMEM) supplemented with 10% heat-inactivated fetal bovine serum (FBS) and antibiotics (100 U/mL penicillin, 100 μg/mL streptomycin) at 37 °C, 5% CO_2.
2. Inoculate 2×10^5 cells/mL in a 12-well plate and incubate for 24 h until cells adhere to the plates.
3. Remove the medium and incubate with Borna disease virus (BoDV) diluted in serum-free medium for 1 h at 37 °C, 5% CO_2 (*see* **Note 3**).

4. Replace virus-containing, serum-free medium with virus-free, serum-containing medium.
5. Culture for several hours or days at 37 °C, 5% CO_2.
6. Remove medium, add 350 μL RL buffer (FastGene RNA Premium Kit) supplemented with 1% 2-mercaptoethanol.
7. Collect the cell lysate into a microtube and vortex intensely for 1 min.
8. Extract total RNA, including small RNAs, using the FastGene RNA Premium Kit with FastGene miRNA enhancer.
9. Measure the quantity of purified total RNA using Nanodrop One.
10. Use the extracted RNA for miRNA cDNA synthesis with specific stem-loop RT primers [22] and the High-Capacity cDNA Reverse Transcription Kit (Fig. 1).
11. Use the extracted RNA for β-Tubulin cDNA synthesis with random hexamer primers and the High-Capacity cDNA Reverse Transcription Kit.
12. Perform qRT-PCR using the KAPA SYBR Fast qPCR kit, specific primers for each gene, and QuantStudio 3 Real-Time PCR system, with the provided protocol (Fig. 2).
13. Normalize miRNA expression levels by β-Tubulin expression levels (Fig. 3) (*see* **Note 4**).

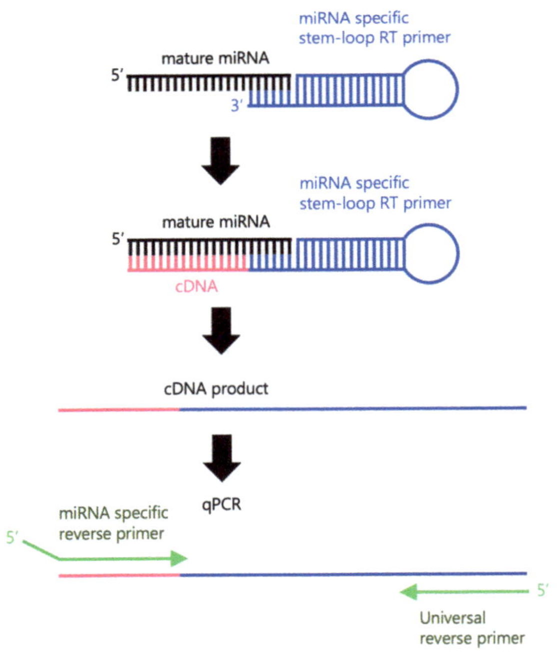

(i) Primer + template mix

Component	Volume
Total RNA	4 μl (0.5 μg)
Step-loop RT primer (0.5 μM)	1 μl
	5 μl

→ 65°C, 5 min

(ii) Enzyme mix

Component	Volume
10 × RT buffer	1 μl
25 × dNTP mix	0.4 μl
RTranscriptase	0.5 μl
RNase free water	3.1 μl
	5 μl

→ (i) + (ii) = 10 μl

PCR protocol
25°C, 10 min → 37°C, 120 min → 85°C, 5 min → 4°C, ∞

Fig. 1 RT protocol for mature miRNA using a stem-loop RT primer

(a) Primer mix (/well)

Component	Volume
F primer (100 μM)	0.025 μl
R primer (100 μM)	0.025 μl
DNase free water	1.95 μl
	2 μl

(b) Enzyme + cDNA mix (/well)

Component	Volume
KAPA SYBR Fast qPCR mix	5 μl
cDNA	2.8 μl
DNase free water	0.2 μl
	8 μl

→ (a) + (b) = 10 μl / well in 96-well PCR plate

qPCR protocol

95°C, 3 min → 95°C, 5 sec→ 60°C, 1 min → 4°C, ∞

40 cycle

Fig. 2 qPCR protocol using SYBR Green reagent

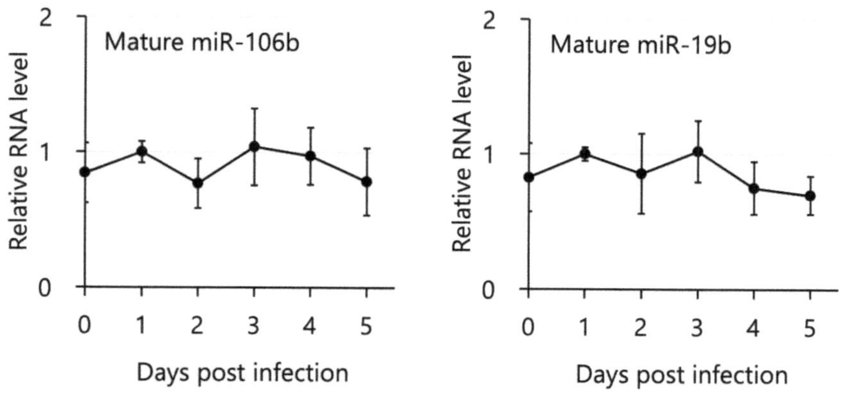

Fig. 3 Quantification of mature miR-106b and miR-19b following BoDV infection

3.2 Target RNA Prediction Using Web Tool

After determining miRNA expression levels, the subsequent step involves predicting miRNA and target RNA interactions. While a basic method is computationally searching for genes with complementary sequences to the miRNA seed sequence in the 3′UTR, a more convenient and accurate approach is to utilize web tools such as TargetScan [15, 16, 23, 24] (*see* **Note 5**).

1. Open "https://www.targetscan.org/vert_80/" with a web browser.
2. Select the species as "Human."
3. Enter the microRNA name, e.g., "miR-106b-5p" (*see* **Note 6**).
4. Click "Submit."

5. Download the predicted target gene list by clicking "Download table."
6. Choose the desired format (txt or xlsx) based on your purpose.

3.3 Experimental Validation of Target RNA Repression by miRNA

To experimentally validate interactions between miRNAs and predicted target genes obtained through web tools, follow the steps outlined below (*see* **Note 7**). Subsequently, elucidate the biological function based on the validated miRNA interaction and its expression level (*see* **Note 8**).

1. Construct two types of psiCHECK-1 plasmids encoding the *Renilla* luciferase gene fused with the 3′UTR of the target gene: a wild-type (WT) plasmid and a mutated plasmid with alterations in the miRNA target sites to prevent miRNA-target RNA interaction (Fig. 4) (*see* **Note 9**).
2. Inoculate 1×10^5 cells/mL of a cell suspension into 24-well plates 1 day before transfection.
3. Transfect cells with psiCHECK-1 plasmids encoding *Renilla* luciferase and the 3′UTR of the target gene (0.25 μg) and pGL3-control plasmid encoding firefly luciferase (0.25 μg) using transfection reagents such as PEI or Lipofectamine.
4. Culture cells for 24 h post-transfection at 37 °C, 5% CO_2.

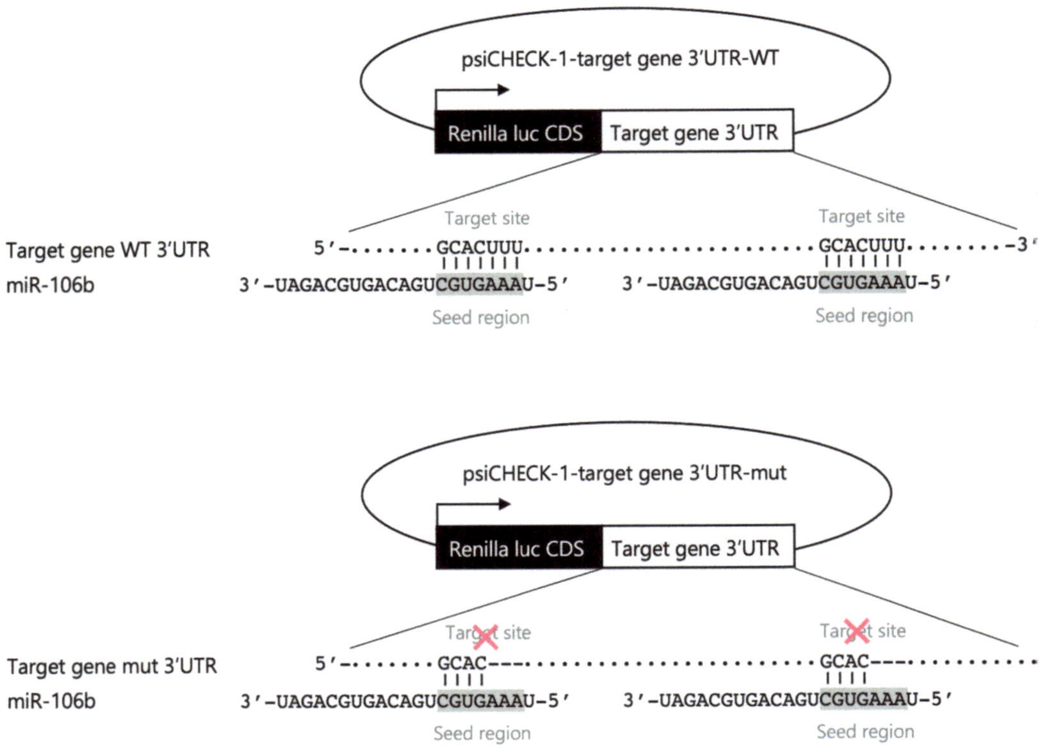

Fig. 4 Plasmids for validating miRNA and target gene interactions

5. Remove the medium and add 250 μL Passive Lysis Buffer to each well.
6. Gently shake plates for 15 min at room temperature.
7. Measure luciferase activity using a dual-luciferase reporter assay system.
8. Calculate relative luciferase activity (Firefly luciferase/*Renilla* luciferase) (*see* **Note 10**).

4 Notes

1. For comprehensive analysis of miRNA expression, small RNA-seq is optimal. Specific miRNA expression levels can be measured by qRT-PCR when individual assessment is required.
2. The effectiveness of inhibiting or introducing miRNA depends on its expression levels. Abundant miRNA expression favors inhibition, while low expression levels warrant miRNA introduction to investigate its function.
3. While viral infection is employed in this section, alternative treatments such as cytokines or inhibitors can be used based on experimental goals.
4. While β-Tubulin is used for normalization in this section, other housekeeping genes like actin or GAPDH can be considered. Ensure these genes' expression remains stable after treatments.
5. Various web tools are available, with the choice dependent on specific research goals (Table 1). Algorithms often consider factors like sequence complementarity, evolutionary conservation, secondary structure, thermodynamic stability, and the number of target sites.
6. Specify whether the investigated miRNA strand is "5p" or "3p" since they possess different seed sequences derived from pre-miRNAs.
7. Consider mutagenesis for specific miRNA-gene interactions, as described above, and employ RNA-seq to assess broader effects after introducing miRNA mimics or inhibitors.
8. Various experimental approaches are available to validate miRNA-target gene interactions based on the expression levels of the miRNA of interest. Use CRISPR knockout or miRNA inhibitors to reduce miRNA activity when the expression level is abundant, and employ miRNA mimics to enhance miRNA effects when the expression level is not abundant.
9. When predicted target genes have multiple target sites, introduce mutations into all sites for accurate validation.

Table 1
List of miRNA-target prediction web tool and their characteristics

Name	URL	Characteristics	References
TargetScan	https://www.targetscan.org/vert_80/	Ranked based on the predicted efficacy of targets and evolutionary conservation	[23, 15, 16, 24]
DIANA-microT	https://dianalab.e-ce.uth.gr/microt_webserver/#/	Searching target sites on CDS and 3′UTR; searching multiple miRNA–mRNA by adding their names; possible to combine predictions with experimental support (DIANA-TarBase)	[25, 26]
TargetMiner	https://www.isical.ac.in/~bioinfo_miu/targetminer20.htm	Searching multiple miRNA–mRNA simultaneously by uploading files	[27]
PITA	https://genie.weizmann.ac.il/pubs/mir07/	Accessibility of the target site	[28]
RNA22	https://cm.jefferson.edu/rna22/	Possible to select the type of targeted RNA from mRNA, lincRNA, and lncRNA	[29]
RNAhybrid	https://bibiserv.cebitec.uni-bielefeld.de/rnahybrid	Minimum free energy for miRNA–mRNA	[30]

10. An upregulation in relative luciferase activity in mutated plasmids compared to WT plasmids suggests that endogenous miRNA targets the predicted genes through mutated target sites.

Acknowledgments

I appreciate Yuka Kaneko MSc for technical support. This work was supported by the Canon Foundation and the MSD Life Science Foundation to TT.

References

1. Bartel DP (2004) MicroRNAs: genomics, biogenesis, mechanism, and function. Cell 116: 281–297
2. Wilson RC, Doudna JA (2013) Molecular mechanisms of RNA interference. Annu Rev Biophys 42:217–239
3. Fire A, Xu S, Montgomery MK et al (1998) Potent and specific genetic interference by double-stranded RNA in Caenorhabditis elegans. Nature 391:806–811
4. Kozomara A, Griffiths-Jones S (2014) miRBase: annotating high confidence microRNAs using deep sequencing data. Nucleic Acids Res 42:D68–D73
5. Lee Y, Ahn C, Han J et al (2003) The nuclear RNase III Drosha initiates microRNA processing. Nature 425:415–419
6. Gregory RI, Yan KP, Amuthan G et al (2004) The microprocessor complex mediates the genesis of microRNAs. Nature 432:235–240

7. Denli AM, Tops BB, Plasterk RH et al (2004) Processing of primary microRNAs by the microprocessor complex. Nature 432:231–235
8. Lund E, Güttinger S, Calado A et al (2004) Nuclear export of microRNA precursors. Science 303:95–98
9. Doi N, Zenno S, Ueda R et al (2003) Short-interfering-RNA-mediated gene silencing in mammalian cells requires dicer and eIF2C translation initiation factors. Curr Biol 13:41–46
10. Liu J, Carmell MA, Rivas FV et al (2004) Argonaute2 is the catalytic engine of mammalian RNAi. Science 305:1437–1441
11. Meister G, Landthaler M, Patkaniowska A et al (2004) Human Argonaute2 mediates RNA cleavage targeted by miRNAs and siRNAs. Mol Cell 15:185–197
12. Matranga C, Tomari Y, Shin C et al (2005) Passenger-strand cleavage facilitates assembly of siRNA into Ago2-containing RNAi enzyme complexes. Cell 123:607–620
13. Behm-Ansmant I, Rehwinkel J, Doerks T et al (2006) mRNA degradation by miRNAs and GW182 requires both CCR4:NOT deadenylase and DCP1:DCP2 decapping complexes. Genes Dev 20:1885–1898
14. Adams D, Gonzalez-Duarte A, O'Riordan WD et al (2018) Patisiran, an RNAi therapeutic, for hereditary transthyretin amyloidosis. N Engl J Med 379:11–21
15. Lewis BP, Burge CB, Bartel DP et al (2005) Conserved seed pairing, often flanked by adenosines, indicates that thousands of human genes are microRNA targets. Cell 120:15–20
16. Friedman RC, Farh KK, Burge CB et al (2009) Most mammalian mRNAs are conserved targets of microRNAs. Genome Res 19:92–105
17. Hausser J, Syed AP, Bilen B et al (2013) Analysis of CDS-located miRNA target sites suggests that they can effectively inhibit translation. Genome Res 23:604–615
18. Hoffmann B, Tappe D, Höper D et al (2015) A variegated squirrel Bornavirus associated with fatal human encephalitis. N Engl J Med 373:154–162
19. Bode L, Xie P, Dietrich DE et al (2020) Are human Borna disease virus 1 infections zoonotic and fatal? Lancet Infect Dis 20:650–651
20. Kaneko Y, Naito Y, Koide R et al (2023) The regulation of persistent Borna disease virus infection by RNA silencing factors in human cells. Biochem Biophys Res Commun 658:122–127
21. Kaneko Y, Takahashi T (2023) AraC-induced neuron-like differentiation of human NTERA2/D1 cells and quantification of endogenous pre-mir-106b and 19b levels. Micro Publ Biol. https://doi.org/10.17912/micropub.biology.000803
22. Chen C, Ridzon DA, Broomer AJ et al (2005) Real-time quantification of microRNAs by stem-loop RT-PCR. Nucleic Acids Res 33:e179
23. Grimson A, Farh KK, Johnston WK et al (2007) MicroRNA targeting specificity in mammals: determinants beyond seed pairing. Mol Cell 27:91–105
24. Garcia DM, Baek D, Shin C et al (2011) Weak seed-pairing stability and high target-site abundance decrease the proficiency of lsy-6 and other microRNAs. Nat Struct Mol Biol 18:1139–1146
25. Reczko M, Maragkakis M, Alexiou P et al (2012) Functional microRNA targets in protein coding sequences. Bioinformatics 28:771–776
26. Paraskevopoulou MD, Georgakilas G, Kostoulas N et al (2013) DIANA-microT web server v5.0: service integration into miRNA functional analysis workflows. Nucleic Acids Res 41:W169–W173
27. Bandyopadhyay S, Mitra R (2009) TargetMiner: microRNA target prediction with systematic identification of tissue-specific negative examples. Bioinformatics 25:2625–2631
28. Kertesz M, Iovino N, Unnerstall U et al (2007) The role of site accessibility in microRNA target recognition. Nat Genet 39:1278–1284
29. Miranda KC, Huynh T, Tay Y et al (2006) A pattern-based method for the identification of MicroRNA binding sites and their corresponding heteroduplexes. Cell 126:1203–1217
30. Rehmsmeier M, Steffen P, Hochsmann M et al (2004) Fast and effective prediction of microRNA/target duplexes. RNA 10:1507–1517

Chapter 12

Unveiling Functional Long-Range RNA Structures via RIC-seq Analysis

Sergey Margasyuk and Dmitri D. Pervouchine

Abstract

RNA in situ conformation sequencing (RIC-seq) is an assay that probes the conformation and structure of RNA using proximity ligation principle. In this assay, RNA molecules are cross-linked through RNA-binding proteins (RBPs) in vivo, which allows the identification of contacts that occur within physiological RNA–RBP complexes and exist in relevant biological contexts. Here, we describe two bioinformatic pipelines: one that extracts RNA contacts from RIC-seq data (RNAcontacts) and the other that uses the obtained information on RNA contacts to predict stable complementary elements of the RNA structure (PHRIC). The former can be used as a standalone tool to discover RNA–RNA interactions *in cis* and *in trans*, while the latter can be viewed as its extension, which searches for pairs of nested contact clusters and predicts complementary interactions between them. Both pipelines are implemented in Snakemake, a reproducible and scalable workflow management system for rapid and uniform processing of multiple transcriptomic datasets.

Key words RIC-seq, RNA structure, RNA contacts, Free energy, Proximity ligation

1 Introduction

RNA conformation capture methods are based on cross-linking, digestion, religation, and high-throughput sequencing to determine spatial proximity of cellular RNAs [1]. A remarkable progress has been made in exploring different implementations of this technique since the first such method was described in 2011 [2–8]. The common core step of all RNA conformation capture methods is the high-throughput sequencing of hybrids formed by the interacting RNA fragments. An important caveat in mapping these hybrids to a reference genome is that they contain exon–exon junctions resulting from pre-mRNA splicing as well as proximity ligation products, in which RNA conformation is encoded [9]. Therefore, it is of interest to address this caveat because the existing computational tools for short read mapping do not account for two distinct split types.

Michiaki Hamada et al. (eds.), *RNA-RNA and RNA-DNA Interactions*, Methods in Molecular Biology, vol. 2949,
https://doi.org/10.1007/978-1-0716-4670-0_12,

Here, we describe two computational pipelines for the analysis of RNA conformation capture data, with a particular focus on the RNA in situ conformation sequencing (RIC-seq) protocol. The first pipeline, called RNAcontacts, serves as a principal tool for mapping short reads with two distinct split types: the canonical intronic GT/AG junctions resulting from pre-mRNA splicing and non-GT/AG splits resulting from RNA proximity ligation [9]. It does so by running the STAR aligner in a two-pass mode [10]. On the first pass, it aligns a sample-matched set of paired-end RNA-seq data, which represents a control experiment without fragments generated by proximity ligation, in order to identify splice junctions that are expressed in a specific biological condition. On the second pass, short reads generated by the RIC-seq experiment are aligned to the reference genome using splice junctions identified on the first pass as bona fide introns. The output of RNAcontacts is the list of RNA contacts, reported either as a list of ligation junctions or as a list of RNA contact clusters, in which closely located junctions are aggregated.

The second pipeline, called PHRIC, is an extension of RNAcontacts. It identifies pairs of stretches of the RNA sequence that are surrounded by bilateral RNA contacts and performs thermodynamic folding of the extracted fragments to predict core RNA structure elements [11]. In application to conserved intronic sequences, such contacts were shown to be associated with remarkable properties of RNA structures such as equilibrium free energy, the presence of compensatory substitutions, and the occurrence of A-to-I RNA editing sites and forked eCLIP peaks [12]. The PHRIC pipeline takes the output of RNAcontacts as input, uses a computationally effective procedure to identify pairs of nested contact clusters (PNCC), extracts their respective nucleotide sequences, and passes them to PREPH, a tool for predicting long, nearly perfect complementary matches in a pair of input sequences [13].

2 Materials

2.1 Hardware

x86–64 Linux workstation or HPC cluster (requires additional Snakemake setup) with at least 45 GB RAM.

2.2 Software

1. Conda installation with Mamba package manager. We use the miniforge installer from https://github.com/conda-forge/miniforge.

```
$ curl -L -O "https://github.com/conda-forge/
miniforge/releases/latest/download/Miniforge3-
$(uname)-$(uname -m).sh"
```

```
$ bash Miniforge3-$(uname)-$(uname -m).sh -b
$ source ~/.bashrc
```

2. Snakemake installation through conda https://github.com/snakemake/snakemake:

```
$ conda activate base
$ mamba create -c conda-forge -c bioconda -n snake-
make snakemake==7.32.4
```

3. RNAcontacts and PHRIC pipelines

```
$ mkdir -p ~/Desktop/lrRNAc/ && cd ~/Desktop/lrRNAc/
$ git clone https://github.com/smargasyuk/RNAcon-
tacts
$ git clone https://github.com/smargasyuk/PHRIC
```

4. sra-toolkit (https://github.com/ncbi/sra-tools), we use conda to install it:

```
$ conda activate base
$ mamba create -c conda-forge -c bioconda -n sratools
sra-tools==3.0.7
```

5. python3 environment for visualization scripts:

```
$ conda activate base
$ mamba create -c conda-forge -c bioconda -n rnac-
plots python==3.11.4 click==8.1.7 pandas==1.5.3
seaborn==0.12.2
```

6. visualization scripts:

```
$ git clone https://github.com/smargasyuk/MiMB-
lrRNAc-scripts.git ~/Desktop/lrRNAc/scripts
```

7. STAR (https://github.com/alexdobin/STAR), fastp (https://github.com/OpenGene/fastp), genomepy (https://github.com/vanheeringen-lab/genomepy), samtools (https://github.com/samtools/samtools), and bedtools (https://

github.com/arq5x/bedtools2) are used; they are installed automatically by Snakemake using Conda.

3 Methods

3.1 Data Preparation

1. Download RIC-seq and RNA-seq reads.

   For demonstration, we chose HeLa RIC-seq data from GEO (GSE127188) and RNA-seq from the same cell line from ENCODE (ENCSR000CPR).

```
$ mkdir -p ~/Desktop/lrRNAc/raw/HeLa/{RIC-seq,RNA-
seq}
$ cd ~/Desktop/lrRNAc/raw/HeLa/RIC-seq
$ conda activate sratools
$ prefetch SRR8632820 && fasterq-dump --split-files
SRR8632820
$ prefetch SRR8632821 && fasterq-dump --split-files
SRR8632821
$ gzip *.fastq
$ cd ~/Desktop/lrRNAc/raw/HeLa/RNA-seq
$ wget https://www.encodeproject.org/files/
ENCFF000FOK/@@download/ENCFF000FOK.fastq.gz
$ wget https://www.encodeproject.org/files/
ENCFF000FOM/@@download/ENCFF000FOM.fastq.gz
$ wget https://www.encodeproject.org/files/
ENCFF000FOV/@@download/ENCFF000FOV.fastq.gz
$ wget https://www.encodeproject.org/files/
ENCFF000FOY/@@download/ENCFF000FOY.fastq.gz
```

2. Configure the pipelines. Edit the sample sheet (file names for HeLa are already set): set the correct absolute paths to the downloaded fastq.gz files; set the "genome" column to "hg38" or the desired genome assembly for all samples. After editing, link the sample sheet to PHRIC:

```
$ nano ~/Desktop/lrRNAc/RNAcontacts/config/samples.
tsv
$ rm ~/Desktop/lrRNAc/PHRIC/config/samples.tsv
$ cat ~/Desktop/lrRNAc/RNAcontacts/config/samples.tsv
| sed 's/hg38/GRCh38/g' > ~/Desktop/lrRNAc/PHRIC/con-
fig/samples.tsv
```

3. Download the GRCh38 genome assembly and GENCODE annotation using genomepy from RNAcontacts:

```
$ cd ~/Desktop/lrRNAc/RNAcontacts/
$ conda activate snakemake
$ snakemake --use-conda -c1 resources/genomes/GRCh38/
GRCh38.annotation.gtf
```

Make the resulting files available to PHRIC and remove everything except the main reference chromosomes from the list:

```
$ mkdir -p ~/Desktop/lrRNAc/PHRIC/resources/genomes/
GRCh38/
$   cat   ~/Desktop/lrRNAc/RNAcontacts/resources/gen-
omes/GRCh38/GRCh38.annotation.gtf  |  gzip  >  ~/Desk-
top/lrRNAc/PHRIC/resources/genomes/GRCh38/annota-
tion.gtf.gz
$ ln -sT ~/Desktop/lrRNAc/RNAcontacts/resources/gen-
omes/GRCh38/GRCh38.fa          ~/Desktop/lrRNAc/PHRIC/
resources/genomes/GRCh38/genome.fa
$ grep -v "_" ~/Desktop/lrRNAc/RNAcontacts/resources/
genomes/GRCh38/GRCh38.fa.sizes  >  ~/Desktop/lrRNAc/
PHRIC/resources/genomes/GRCh38/chromSizes
```

4. Download the repeat track from the UCSC table browser. Select "genome: Human; assembly: Dec.2013 (GRCh38/hg38); group: Repeats; track: RepeatMasker; table: rsmk" in "UCSC Table Browser"; select "BED—Browser Extensible Data" output type; download the resulting file and save it to

```
~/Desktop/lrRNAc/PHRIC/resources/genomes/GRCh38/
RMSK.bed
```

3.2 Extraction of Ligation Points from RIC-Seq Data

To identify RNA contacts, we align short reads with the STAR aligner and extract ligation points from the canonical (BAM) and chimeric (Chimeric.out.junctions) outputs of STAR. To account for the presence of two types of splits, the alignment is performed in two passes. The first pass uses RNA-seq data with a strict penalty for non-GT/AG splits to select for spliceosomal introns; the second pass uses RIC-seq data with a relaxed penalty for non-GT/AG splits to detect chimeric reads encoding RNA contacts. The alignment of

RNA-seq data on the first pass is performed in the paired-end (PE) mode, while the alignment on the second pass is performed in the single-end mode because RIC-seq data may contain chimeric junctions at arbitrary genomic distances or *in trans*. The ligation points are then extracted from output files, and the intronic junctions from the first pass are removed from the output.

RNAcontacts implements these steps in the Snakemake workflow language. It consists of the following steps:

1. Trimming of adapters with `fastp`.
2. First-pass STAR alignment of RNA-seq data with default parameters in PE mode.
3. Second-pass STAR alignment of RIC-seq data in SE mode with the relaxed parameters: '—chimOutType Junctions —chimSegmentMin 15 —chimJunctionOverhangMin 15 —chimScoreJunctionNonGTAG −1 —scoreGapNoncan −1 —scoreGapATAC -1 —scoreGapGCAG −1 —chimSegmentReadGapMax 3 —outFilterMatchNminOverLread 0.5 —outFilterScoreMinOverLread 0.5.'
4. Extraction of junction (split read) coordinates from the collinear (BAM) and chimeric outputs of STAR. For multi-mapped reads, junction coordinates are extracted from the primary alignment only.

The rationale behind these options is as follows. The parameters *—chimSegmentMin 15 —chimJunctionOverhangMin 15* enable search for chimeric segments of at least 15 nt, while the second option requires the exon length next to the chimeric junction to be at least 15 nt. The *—chimSegmentReadGapMax 3* accounts for the additional biotinylated cytosine residue in RIC-seq protocol [10]. The penalty score for all non-canonical splits is reduced to −1 to detect all possible RNA hybrids (not only GT/AG). Users can adjust these settings to accommodate their specific needs.

To launch the pipeline (change resource limits to the appropriate ones for your system), run the following code:

```
$ cd ~/Desktop/lrRNAc/RNAcontacts
$ conda activate snakemake
$  snakemake  --use-conda  --cores  8  --resources='-
mem_mb=40000' junctions
```

This step generates files, which contain junctions extracted from the collinear and chimeric outputs, and places them in ~/Desktop/lrRNAc/RNAcontacts/results/hg38/HeLa/junctions/. To plot the distributions of spreads (i.e., the distance

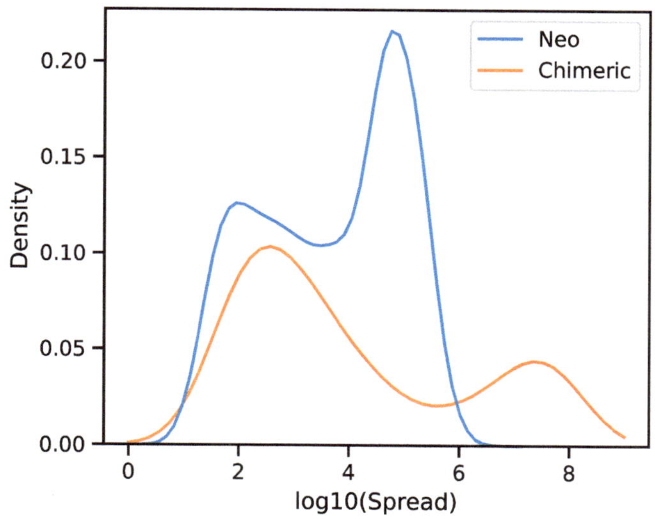

Fig. 1 The distribution of spreads (i.e., the distance between contact points) for collinear and chimeric output of STAR

between contact points) for intrachromosomal contacts, merge the tables together and run the visualization script:

```
$ cd ~/Desktop/lrRNAc/RNAcontacts
$ mkdir -p data/tables ~/Desktop/lrRNAc/plots/
$ zcat results/hg38/HeLa/junctions/RIC-seq_HeLa_rRNA_de-
pleted_rep*/Neo.tsv.gz | awk -v 'OFS=\t' '{print
$0,"Neo"}' > data/tables/junctions.tsv
$ zcat results/hg38/HeLa/junctions/RIC-seq_HeLa_rRNA_de-
pleted_rep*/Chimeric.tsv.gz | awk -v 'OFS=\t' '{print
$0,"Chimeric"}' >> data/tables/junctions.tsv
$ conda activate rnac-plots
$ python ~/Desktop/lrRNAc/scripts/jxn_spread_hist.py -i
data/tables/junctions.tsv -o ~/Desktop/lrRNAc/plots/
jxn_spread_hist.pdf
```

Figure 1 exemplifies the plot generated by this command for the HeLa dataset. The term "neo-junctions" refers to the contact points extracted from BAM output. The term "chimeric" refers to the contact points extracted from chimeric alignments (ones placed on the same strand on the same chromosome by STAR). Both distributions have two modes, with the lower mode at ~1000 nt, which corresponds to intragenic contacts encoded by both neo-junctions and chimeric splits. The second mode for neo-junctions is due to the upper limit of 250,000 nt split imposed by the STAR aligner on collinear splits. Longer contacts *in cis* correspond to the second mode of the chimeric distribution.

In proximity ligation methods, different copies of the same RNA can be digested and religated stochastically, thus resulting in considerable variability in the coordinates of ligation junctions [9]. Even reading the same sequence from the positive or the negative strand may produce different split coordinates due to the lack of consensus sequences that characterize the split. Considering this technical and biological variation, we introduced an optional step to aggregate the ligation junctions using single linkage clustering with the default distance cutoff of 10 nt.

To obtain RNA clusters, run the following code:

```
$ cd ~/Desktop/lrRNAc/RNAcontacts
$ conda activate snakemake
$ snakemake --use-conda --cores 8 --resources='-
mem_mb=40000' contacts
```

The output of this step consists of RNA contact clusters for the collinear and chimeric junctions, which are placed in ~/Desktop/lrRNAc/RNAcontacts/results/hg38/HeLa/contacts/. The resulting RNA contact clusters are characterized by the coordinates of the corresponding clusters and the sum of read counts corresponding to their constituent ligation junctions.

3.3 Search for Pairs of Nested Contact Clusters and Prediction of Complementarity Between Them

This section describes the PHRIC pipeline, which identifies RNA secondary structures that are supported by RIC-seq contacts [11]. Toward this goal, it first extracts intragenic contacts (i.e., contacts between RNA bases in the same gene) from the list of contacts produced at Subheading 3.2 and performs their clustering. Next, it selects pairs of nested contact clusters (PNCC) with distances 10–50 nt between the contacting regions, extracts the nucleotide sequences corresponding to the intervals enclosed by the contacts, and predicts RNA secondary structure formed by the extracted intervals using the PREPH tool [13]. This procedure results in a list of candidate RNA secondary structures that are characterized by the free energy of hybridization.

The PHRIC pipeline is also implemented in Snakemake. It consists of the following steps:

1. Extraction of intrachromosomal junctions from RNAcontacts output.
2. Clustering of contacts. The single-linkage clustering with a maximum distance of 10 is performed for both left and right ligation points of the contacts. Each contact cluster is characterized by the total number of supporting RIC-seq reads.
3. Selection of intragenic contacts.

4. Search for PNCC with a distance of 10–50 nt between both contacting regions.
5. Filtering of PNCC by read support of at least two for both contact clusters; extraction of regions between contacting clusters (referred to as folding regions).
6. Removal of folding regions intersecting genomic repeats.
7. Prediction of complementarity between folding regions using PREPH.

Make the RNAcontacts results available for PHRIC and launch it with the following commands (change resource limits to the appropriate ones for your system):

```
$ ln -sT ~/Desktop/lrRNAc/RNAcontacts/results/hg38 ~/
Desktop/lrRNAc/PHRIC/resources/RNAcontacts-output/GRCh38
$ cd ~/Desktop/lrRNAc/PHRIC/
$ conda activate snakemake
$     snakemake     --use-conda     --cores     8     --
resources='mem_mb=40000'
```

This pipeline generates a table with candidate RNA structures in TSV format (~/Desktop/lrRNAc/PHRIC/results/GRCh38/S16.tsv) and a BED file for visualization (~/Desktop/lrRNAc/PHRIC/results/GRCh38/S16.bed). We recommend considering RNA structures with the equilibrium free energy (ΔG) of at least 15 kcal/mol by absolute value. The following script visualizes the distribution of |ΔG| (Fig. 2):

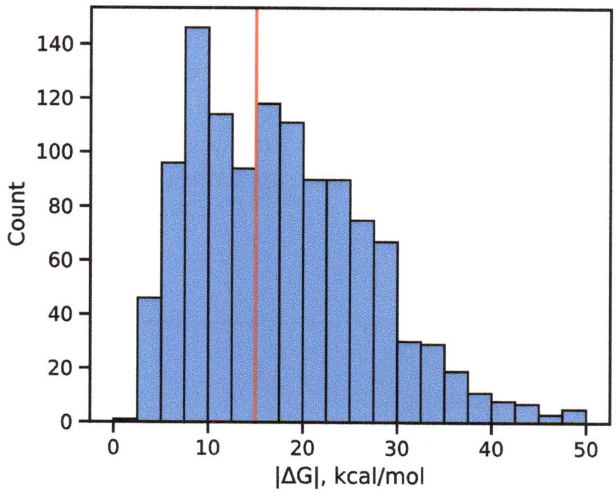

Fig. 2 The distribution of equilibrium free energies (|ΔG|) for the predicted structures. The red line shows the recommended cutoff of 15 kcal/mol

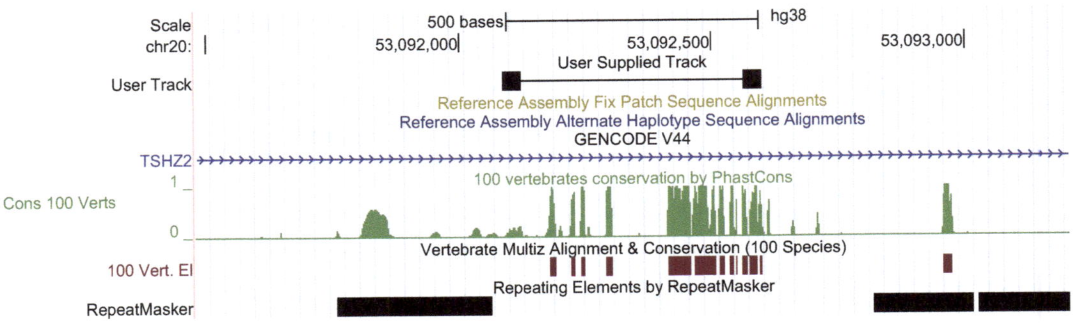

Fig. 3 Visualization of the predicted RNA structure in the human TSHZ2 gene. The pair of complementary regions is indicated by black boxes in the `PHRIC` prediction track

```
$ conda activate rnac-plots
$ python ~/Desktop/lrRNAc/scripts/struct_energy_hist.py
-i results/GRCh38/S16.tsv -o ~/Desktop/lrRNAc/plots/
struct_energy_hist.pdf
```

The predicted structures can be visualized in the UCSC Genome Browser or using other visualization software. For UCSC Genome Browser, select the first 12 columns of the BED file:

```
$ mkdir -p data/tables
$ echo 'track name=PHRIC description="PHRIC predic-
tions"' > data/tables/S16.bed
$ cut -f1-12 results/GRCh38/S16.bed >> data/tables/S16.
bed
```

The output file can be uploaded to the UCSC Genome Browser as a custom track. Figure 3 exemplifies the RNA structure predicted for the human TSHZ2 gene. We recommend prioritizing RNA structures located in conserved intronic regions (e.g., selected by PhastCons from multiple alignments).

4 Notes

The described principle of two-pass mapping represents a universal solution to the problem of aligning short reads with two distinct split types that are generated by RNA conformation capture assays, including the RIC-seq experiments. It can be implemented not only using STAR mapper but also using any short read aligner that takes into account split reads. Users can choose to run the automatic pipeline presented here or implement their own workflow with additional filters on read quality, types of chimeric alignments, the maximum unmapped read sequence between two chimeric segments to account for the addition of biotinylated cytosine residues, etc. Clustering of contacts in RNAcontacts pipeline is an optional step that helps to reduce the amount of information in the output and also allows for prioritization of contacting clusters based on the number of supporting split reads. The assessment of statistical significance of the observed RNA contacts, however, remains a challenge due to the inherent sparsity of RIC-seq data at single-base resolution. Adapter trimming appears to be an essential step since untrimmed sequences can result in a large number of false-positive RNA contacts.

RIC-seq experiments provide a snapshot of cellular RNA conformation in complexes with RNA-binding proteins and other molecules. Part of this global picture is RNA structure, which consists of base-paired complementary regions. Not all RNA contacts identified in RIC-seq experiments necessarily correspond to RNA structure. However, a specific pattern of contacts around double-stranded RNA fragments can be seen in many cases that were studied experimentally [12]. It consists of RNA contacts located outside and inside the base-paired region (Fig. 1 in Ref. [11] and Fig. 1 in Ref. [12]), thus offering a powerful principle of identification of double-stranded RNA fragments. This principle is based on the concept of pairs of nested contact clusters, which can be extracted from the contact data in linear time. However, it should be noted that not all double-stranded regions must necessarily be bilaterally surrounded by RNA contacts, and not all pairs of nested contact clusters necessarily correspond to base-paired regions. Therefore, the assessment of false-positive and false-negative prediction rates in PHRIC remains a future goal. The workflow presented here can be modified, for instance, to fold RNA fragments adjacent to the observed contacts only from one side, to take into account additional information on RNA conformation such as SHAPE-seq [14], or to use alternative computational tools such as RNAplex to predict base pairings [15]. In future studies, the RNA contacts observed in a RIC-seq experiment could be used to predict global RNA structure, including secondary and tertiary structures, as well as interactions with RNA-binding proteins. However, considering inherent stochasticity of RNA proximity ligation assays, the lack of models for RNA–protein interactions, and highly dynamic changes that occur to RNA structure during all steps of intracellular processing, in PHRIC we resort to discovering only bilaterally supported long-range stem-like structures, which appears to be the most reasonable goal at the current state of the art.

Acknowledgments

This work was supported by the Ministry of Science and Education of the Russian Federation (Grant No. 075-10-2021-116).

References

1. Kudla G, Wan Y, Helwak A (2020) RNA conformation capture by proximity ligation. Annu Rev Genomics Hum Genet 21:81–100
2. Kudla G, Granneman S, Hahn D, Beggs JD, Tollervey D (2011) Cross-linking, ligation, and sequencing of hybrids reveals RNA-RNA interactions in yeast. Proc Natl Acad Sci USA 108(24):10010–10015
3. Lu Z, Zhang QC, Lee B, Flynn et al (2016) RNA duplex map in living cells reveals higher-order transcriptome structure. Cell 165(5): 1267–1279
4. Sharma E, Sterne-Weiler T, O'Hanlon D et al (2016) Global mapping of human RNA-RNA interactions. Mol Cell 62(4):618–626
5. Aw JG, Shen Y, Wilm A et al (2016) Vivo mapping of eukaryotic RNA interactomes reveals principles of higher-order organization and regulation. Mol Cell 62(4):603–617
6. Ziv O, Gabryelska MM, Lun AT et al (2018) COMRADES determines in vivo RNA structures and interactions. Nat Methods 15(10): 785–788
7. Cai Z, Cao C, Ji L et al (2020) RIC-seq for global in situ profiling of RNA-RNA spatial interactions. Nature 582(7812):432–437
8. Ye R, Hu N, Xue Y (2023) CRIC-seq protocol for in situ profiling of proximal RNA-RNA contacts associated with RNA-binding proteins. STAR Protoc 4(3):102401
9. Margasyuk SD, Vlasenok MA, Li G et al (2023) RNAcontacts: a pipeline for predicting contacts from RNA proximity ligation assays. Acta Nat 15(1):51–57
10. Dobin A, Davis CA, Schlesinger F et al (2013) STAR: ultrafast universal RNA-seq aligner. Bioinformatics 29(1):15–21
11. Margasyuk S, Zavileyskiy L, Cao C et al (2023) Long-range RNA structures in the human transcriptome beyond evolutionarily conserved regions. PeerJ 11:e16414
12. Margasyuk S, Kalinina M, Petrova M et al (2023) RNA in situ conformation sequencing reveals novel long-range RNA structures with impact on splicing. RNA 29(9):1423–1436
13. Kalmykova S, Kalinina M, Denisov S et al (2021) Conserved long-range base pairings are associated with pre-mRNA processing of human genes. Nat Commun 12(1):2300
14. Loughrey D, Watters KE, Settle AH et al (2014) SHAPE-Seq 2.0: systematic optimization and extension of high-throughput chemical probing of RNA secondary structure with next generation sequencing. Nucleic Acids Res 42(21):e165
15. Tafer H, Hofacker IL (2008) RNAplex: a fast tool for RNA-RNA interaction search. Bioinformatics 24(22):2657–2663

Chapter 13

Tethered Function Assay for Analyzing PIWI-piRNA-Mediated Co-transcriptional Silencing

Kensaku Murano and Yuka W. Iwasaki

Abstract

PIWI-interacting RNAs (piRNAs) form complexes with a series of PIWI proteins to silence their target transposons. In the nucleus, piRNAs associated with *Drosophila* Piwi protein induce co-transcriptional silencing. To understand this regulation, reporter-based forced tethering of Piwi-associated proteins has been utilized as a powerful tool. Here, we describe the use of luciferase reporter assays combined with forced tethering of Piwi-interacting proteins for investigation of the molecular mechanism of Piwi-directed co-transcriptional silencing in *Drosophila*. Here, ovarian somatic cells (OSCs) are used as an example for the construction of this reporter system, but this methodology can also be applied to in vivo analysis using *Drosophila* ovaries.

Key words PIWI-piRNA, Co-transcriptional silencing, Reporter assay, Ovarian somatic cell, *Drosophila*

1 Introduction

PIWI-interacting RNAs (piRNAs) constitute a class of small non-coding RNAs pivotal for safeguarding the genomic integrity of diverse organisms, including *Drosophila melanogaster* [1, 2]. The PIWI-piRNA pathway intricately regulates transposons, mobile genetic elements capable of significantly impacting genomic stability, especially in germline cells. Among the three PIWI subfamily proteins in *Drosophila*, piRNAs associated with the Piwi protein (Piwi-piRNAs) participate in co-transcriptional regulation of target transposons by inducing heterochromatin formation. This regulatory mechanism appears to modulate the expression of target transposons through the interplay between Piwi-piRNAs and various factors associated with nascent RNA transcripts of these transposons (Fig. 1a).

Given that in many cases, Piwi-piRNA-associated silencing factors have unclear roles, an effective assay system is crucial for

Michiaki Hamada et al. (eds.), *RNA-RNA and RNA-DNA Interactions*, Methods in Molecular Biology, vol. 2949, https://doi.org/10.1007/978-1-0716-4670-0_13,

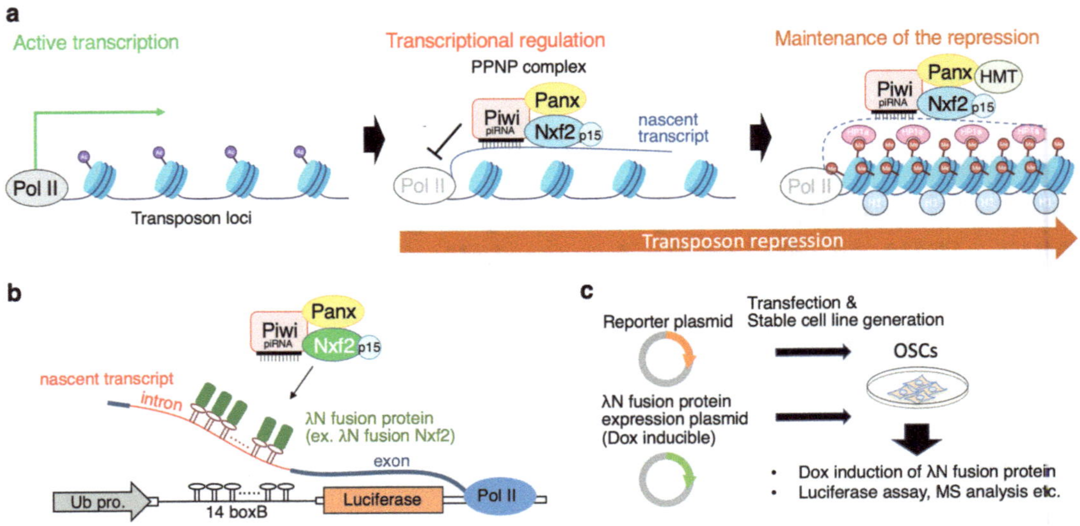

Fig. 1 Piwi-piRNA-mediated co-transcriptional regulation and the tethering reporter system. **a** Sequential regulation of target transposons via the Piwi-piRNA silencing complex. The *Drosophila* Piwi-piRNA complex associates with various factors to initiate silencing. In this illustration, the silencing effector complex, known as the PPNP (also known as PICTS, SFiNX, Pandas) complex, demonstrates stepwise regulation of co-transcriptional silencing. **b** Tethering of factors involved in co-transcriptional silencing to the luciferase reporter through the λN-boxB tethering system artificially induces silencing at the reporter site. Luciferase assays are employed to measure expression levels following the forced tethering of various factors. **c** Workflow of the tethering experiment outlined in this protocol. OSCs are transfected with a luciferase reporter plasmid and a λN fusion protein expression plasmid, followed by the establishment of stable cell lines. These stable cell lines express λN fusion proteins upon induction by Dox, which can be followed by luciferase assays and potentially other related assays to reveal the co-transcriptional silencing effect of different proteins

understanding the mechanisms behind Piwi-piRNA-mediated co-transcriptional regulation. Recently, an approach was implemented to demonstrate that Panx (also known as Silencio) serves as a mediator of transcriptional repression by tethering Panx to the nascent transcript of a reporter [3, 4]. Furthermore, a complex comprising Panx, Nxf2, and p15/Nxt1 has been identified as a silencing effector complex associated with Piwi-piRNAs during co-transcriptional silencing [5–8]. In these studies, a tethering assay of Nxf2 played a crucial role in revealing the hierarchy and ability of the complex to regulate target transposons (Fig. 1b). By examining different time points during the induction of silencing factors in OSCs, these studies unveiled the sequential regulation and nuclear architectural changes occurring during Piwi-piRNA-mediated silencing [7, 9]. Additional studies have demonstrated the applicability of this method in identifying the molecular mechanisms by which novel factors participate in Piwi-piRNA-mediated co-transcriptional silencing [10, 11], further reinforcing the utility of this approach.

Here, we outline a detailed protocol for executing the tethering assay, with a particular emphasis on its application in OSCs (Fig. 1c). The protocol comprises several steps, starting with the preparation of a luciferase reporter plasmid with intronic boxB sequences, which enable the forced tethering of λN fusion proteins [12]. Subsequently, an expression plasmid for producing λN fusion proteins is generated, followed by the establishment of stable OSC lines carrying these expression units. These cell lines are then used to induce co-transcriptional repression by λN fusion proteins through the addition of doxycycline (Dox). Using this system, an extensive array of experimental analyses can be employed to reveal the mechanism of co-transcriptional repression upon the induction of forced tethering. Additionally, this method can also be used to analyze in vivo effects in *Drosophila* ovaries, as previously reported [3–8, 11]. This experimental setup offers a versatile platform for investigating the intricate interplay between Piwi-piRNAs and the transcriptional repression machinery, thereby providing valuable insights into the mechanisms regulating gene expression during germ cell development.

2 Materials

2.1 Construction of Reporter Plasmids

1. QIAquick PCR Purification Kit (Qiagen).
2. SalI and XhoI restriction enzymes.
3. Ligation High (TOYOBO).
4. QIAprep Spin Miniprep Kit (Qiagen).
5. NucleoBond Xtra Midi (TAKARA).
6. Competent *Escherichia coli* strain (DH5α or MACH1).
7. Reporter plasmid (pUb-Luc; *see* **Note 1**).
8. Plasmid harboring 5× boxB sites (BoxB in pcDNA3, Addgene plasmid #29727, *see* **Note 2**).

2.2 OSC Cell Culture

Materials required for culturing OSCs are described in detail in a previous study [13].

2.3 Transfection and Single Colony Isolation

1. Xfect transfection reagent (Clontech).
2. pAc2-Blast (*see* **Note 1**).
3. Blasticidin (InvivoGen).
4. Puromycin dihydrochloride (SIGMA).
5. Cellbanker 1 (TAKARA).

2.4 Construction of a Plasmid Carrying a λN-tagged Gene

1. PiggyBac donor plasmid (pPB-TRE-lamN-HA-3Flag-EcoRV-Tjen-3G-P2A-puroR, *see* **Note 1**).
2. *Eco*R V restriction enzyme.
3. Primers for amplifying the gene of interest.
4. Helper plasmid (pAcM-HyPBase, *see* **Note 1**).

2.5 Luciferase Assay

1. ONE-Glo EX Luciferase Assay System (Promega, E8110).
2. Glo Lysis Buffer, 1× (Promega, E266A).
3. Doxycycline (Clontech).
4. 96-Well Polypropylene Microplate, White, Round bottom (Thermo Scientific, 267350).
5. Microplate Reader, Cytation 5 (BioTek).
6. Protein Assay Dye Reagent Concentrate (Bio-Rad, 5000006).
7. 96-Well Microplate, Clear, Flat bottom (IWAKI, 3881-096).
8. Microplate Reader, iMark (Bio-Rad).

3 Methods

3.1 Generation of an Intronic boxB Reporter Plasmid

In the pUb-Luc plasmid, the firefly luciferase reporter gene is driven by the ubiquitin promoter, including its first intron (Fig. 2). To obtain the intronic boxB reporter construct, a PCR-amplified 5× boxB fragment carrying SalI and XhoI sites is inserted into an XhoI site within the intron of the ubiquitin promoter. Since SalI and XhoI produce compatible cohesive ends, the insertion results in a re-cleavable ligation product with XhoI downstream of the insertion. By sequentially inserting 5× boxB fragment into the recovered XhoI site, reporters with 10 or 15 copies of boxB can be constructed (Fig. 2, *see* **Note 3**).

1. Digest 1 μg of the pUb-Luc plasmid with XhoI using standard protocols. Add 10 units of Alkaline Phosphatase (Calf Intestinal, CIP) to the reaction and incubate at 37 °C for 1 h. Purify the digested plasmid with the QIAquick PCR Purification Kit according to the manufacturer's instructions.
2. Amplify a 5× boxB fragment carrying SalI and XhoI sites from a plasmid harboring 5 × boxB using standard PCR protocols (*see* **Note 2**) and confirm the amplified PCR product by agarose gel electrophoresis. Purify the PCR product with the QIAquick PCR Purification Kit.
3. Digest the 5× boxB fragment with SalI and XhoI, and purify the digested fragment with the QIAquick PCR Purification Kit.
4. Mix the following reagents according to the manufacturer's instructions (*see* **Note 4**).

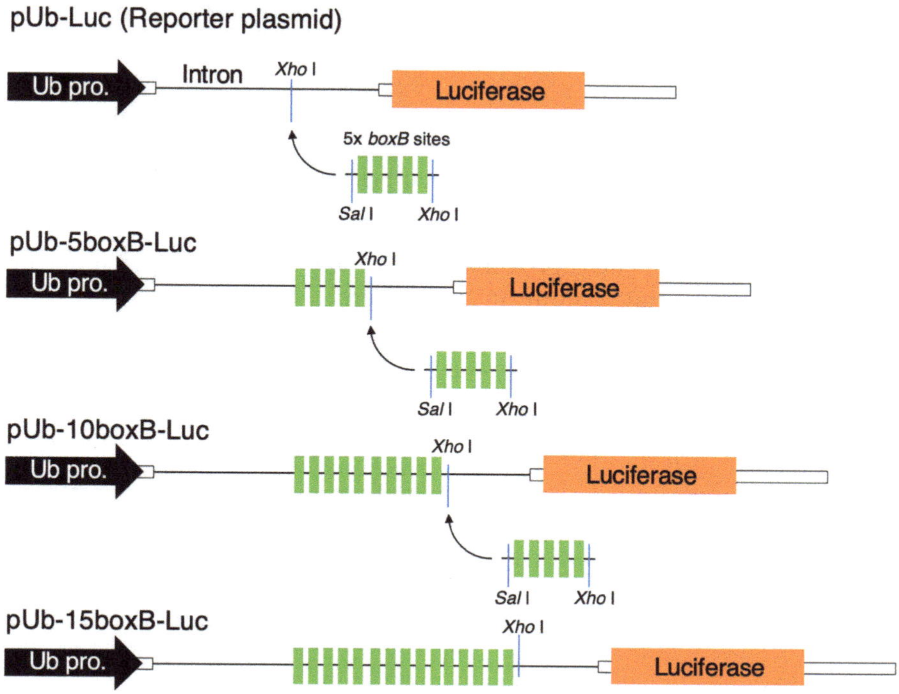

Fig. 2 Construction of reporter plasmids with intronic boxB sequences. The luciferase gene is controlled by the Ubiquitin promoter, including its first intron. The intron contains a site targeted by the XhoI restriction enzyme. To generate a reporter plasmid carrying intronic boxB sequences, a 5× boxB fragment flanked by the SalI site and the XhoI site is inserted into the intronic XhoI site, restoring the XhoI site downstream of the 5× boxB fragment. The ligation product of the cut sites of XhoI and SalI will no longer be recognized by either of the restriction enzymes. Sequential insertion of a 5× boxB fragment at the XhoI site downstream of boxB in the plasmid produces a reporter plasmid with multiples of five copies of intronic boxB sequences (*see* **Note 3**)

The digested pUb-Luc (10 ng/μL)	1 μL
5× boxB fragment (2 ng/μL)	1 μL
Ligation high	2 μL
Total	4 μL

5. Incubate the sample at 16 °C for 30 min.
6. Transform competent cells of *E. coli* with 2 μL of the ligation product and plate on an ampicillin plate.
7. Pick colonies from ampicillin plates, isolate plasmids using a plasmid isolation kit (e.g., QIAprep Spin Miniprep Kit), and confirm the sequence of the insertion by Sanger sequencing.
8. Perform large-scale culturing of *E. coli* transformed with the plasmid. Isolate the plasmid using a plasmid isolation kit (e.g., NucleoBond Xtra Midi). Quantify the plasmid concentration

using the Nanodrop spectrophotometer and adjust the concentration to 1 μg/μL using TE buffer.

3.2 Transfection of Plasmids into OSCs

OSCs are cultured as previously described [13]. Briefly, OSCs were cultured in Shield and Sang M3 Insect Medium supplemented with 10% fly extract, 10% fetal bovine serum, 0.6 mg/mL glutathione, and 10 μg/mL insulin.

1. Plate approximately 1.0×10^6 OSCs on a ϕ35 mm culture dish and culture at 26 °C overnight.
2. Confirm that the cells are approximately 60% confluent at the time of transfection.
3. Mix the following reagents in tubes.

Tube 1

Xfect reaction buffer	75 μL
pUb-14boxB-Luc (1 μg/μL)	2.7 μL
pAc2-Blast (0.3 μg/μL)	1 μL

Tube 2

Xfect reaction buffer	75 μL
Xfect polymer	0.9 μL

4. Mix thoroughly by vortexing for 10 s at high speed.
5. Transfer the solution from Tube 2 to Tube 1.
6. Mix thoroughly by vortexing for 10 s at high speed.
7. Incubate for 10 min at room temperature to allow nanoparticle complexes to form.
8. Remove the OSC culture medium by aspiration, wash the cells twice with 2 mL PBS, and replace the medium with 1.5 mL of Shields and Sang M3 Insect Medium, to remove serum from the cell culture (*see* **Note 5**).
9. Add the entire nanoparticle complex solution dropwise to the cell culture medium. Rock the plate gently back and forth to mix.
10. Incubate OSCs at 26 °C for 3 h.
11. Remove the nanoparticle-containing medium by aspiration and replace the medium with 2 mL of fresh complete OSC culture medium. Incubate OSCs at 26 °C for 48 h.

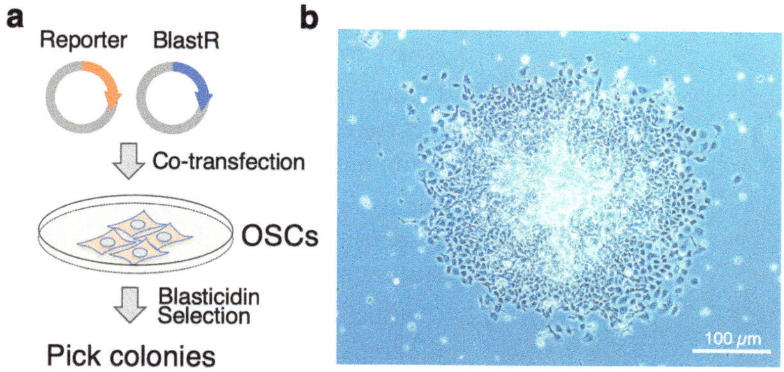

Fig. 3 Isolation of an OSC line carrying the reporter gene. **a** Workflow for generating a stable OSC line carrying the reporter gene. The plasmid encoding a blasticidin resistance gene and the reporter plasmid are co-transfected, followed by blasticidin selection. **b** Example of a well-isolated colony of OSCs after blasticidin selection. This colony, which has grown sufficiently, is suitable for picking

3.3 Isolation of a Stable Cell Line Expressing the Reporter Gene

To generate a cell line expressing the reporter gene, a plasmid encoding a blasticidin-resistance gene and the reporter plasmid are co-transfected, followed by blasticidin selection (Fig. 3a). The gene conferring resistance to blasticidin enables cell survival even in the presence of 50 μg/mL blasticidin for approximately 1 week. Cell death initiates gradually, and the presence of blasticidin is exclusively tolerated by cells that have integrated the blasticidin-resistance gene into their genome. It takes approximately 2 weeks from starting the selection with blasticidin for colonies to reach an optimal size suitable for harvesting (Fig. 3b). Finally, choose cell lines that have incorporated the reporter gene by luciferase assays, as described in Subheading 3.6.

1. Two days after transfection, passage 25% of the cells in 8 mL of OSC culture medium containing 50 μg/mL blasticidin to a ϕ100 mm culture dish.
2. Culture OSCs for 48 h at 26 °C.
3. Replate 25% of the cells in 8 mL of OSC culture medium containing 50 μg/mL blasticidin on a ϕ100 mm culture dish.
4. Culture OSCs for 48 h at 26 °C.
5. Plate 2.0 × 10^6 OSCs into 8 mL of OSC culture medium containing 50 μg/mL blasticidin on a ϕ100 mm culture dish (*see* **Note 6**).
6. Culture OSCs for 48 h at 26 °C.
7. Replace the culture medium with fresh OSC medium containing 5 μg/mL blasticidin.

8. Culture OSCs for at least 5 days until colonies reach a size that is easy to pick (Fig. 3b).
9. Under the microscope, scrape a well-isolated colony with a pipette tip onto a P20 pipette and transfer the cells to a tube containing 30 μL of 0.05% (w/v) Trypsin-EDTA solution.
10. Incubate the tube at 37 °C for 2 min and transfer the cell suspension to a well of a 24-well plate with 500 μL OSC culture medium containing 5 μg/mL blasticidin.
11. Culture OSCs for 3–4 days. Replace the medium with fresh medium containing 5 μg/mL blasticidin every 2 days.
12. Select cell lines with luciferase activity using the method described in Subheading 3.6.
13. Expand positive clones. OSCs can be stored in liquid nitrogen using Cellbanker 1 cryopreservation medium, following the manufacturer's instructions.

3.4 Generation of a Plasmid Carrying a λN-tagged Gene (pPB-TRE-λN-GOI)

The expression vector for producing a λN-tagged protein contains two transcription units flanked by 5′ and 3′ piggyBac ITR (Fig. 4a). In the first transcription unit, the λN-tagged gene is driven by the Tetracycline Response Element (TRE) promoter upon Dox induction. The second transcription unit encodes a reverse tetracycline trans-activator (rtTA) and a puromycin resistance gene, which are bicistronically expressed and separated by the P2A sequence. To generate an expression vector carrying a λN-tagged gene of interest (GOI), the DNA fragment is inserted in frame with the λN-HA-3Flag at the EcoRV site on the PiggyBac donor plasmid (pPB-TRE-lamN-HA-3Flag-EcoRV-Tjen-3G-P2A-puroR (6.8 kbp); Fig. 4a, *see* **Note 1**).

1. Digest the piggyBac donor plasmid with EcoRV restriction enzyme using standard protocols, and purify the digested plasmid with the QIAquick PCR Purification Kit.
2. Amplify the gene of interest using standard PCR protocols (*see* **Note 7**), confirm the amplified PCR product by agarose gel electrophoresis and purify it with the QIAquick PCR Purification Kit.
3. Mix the following reagents in a PCR tube.

Vector fragment (10 ng/μL, 6.8 kbp)	1 μL
Insert fragment	1 μL (=5 fmol, *see* **Note 8**)
HiFi DNA Assembly Master Mix	2 μL
Total	4 μL

4. Incubate the sample in a thermal cycler at 50 °C for 15 min.

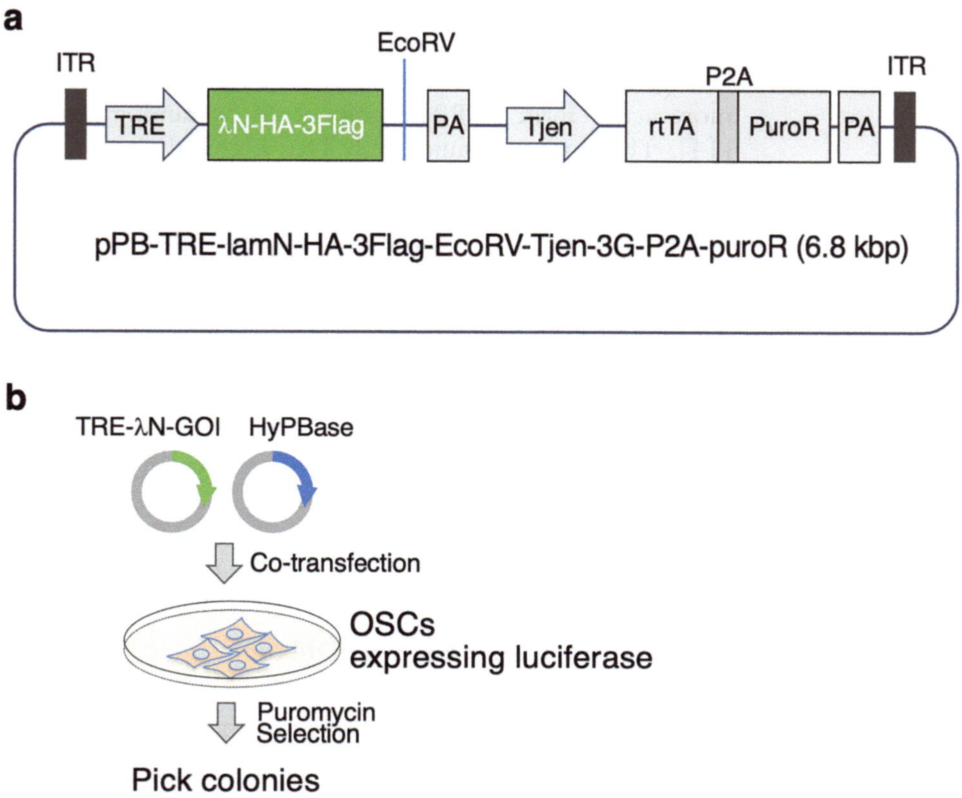

Fig. 4 Establishment of an OSC line carrying a Dox-inducible λN-tagged gene. **a** PiggyBac donor plasmid for the generation of a λN-tagged GOI driven by the TRE promoter. The gene will be inserted into the EcoRV site of the plasmid. ITR: inverted terminal repeat for the piggyBac transposase, PA: SV40 polyA signal, Tjen: enhancer of the tj gene and *Drosophila* synthetic core promoter, rtTA: tetracycline trans-activator, P2A: self-cleaving peptide sequence for bicistronic expression, PuroR: puromycin-resistance gene. **b** Workflow for generating a stable OSC line carrying a Dox-inducible λN-tagged GOI. The donor plasmid encoding a λN-tagged GOI and a plasmid expressing HyPBase are co-transfected, followed by puromycin selection in the presence of blasticidin. Blasticidin is required for the maintenance of the boxB-luciferase reporter gene

5. Transform *E. coli* with the assembled product and plate on an ampicillin plate.
6. Pick colonies from the ampicillin plate, isolate plasmids using a plasmid isolation kit (e.g., QIAprep Spin Miniprep Kit), and confirm the sequence of the insertion by Sanger sequencing.
7. Perform large-scale culturing of *E. coli* transformed with pPB-TRE-λN-GOI. Isolate the plasmid using a plasmid isolation kit (e.g., NucleoBond Xtra Midi). Quantify the plasmid concentration using the Nanodrop spectrophotometer and adjust the concentration to 1 μg/μL using TE buffer.

3.5 Transfection of Plasmids for the piggyBac System into OSCs

The piggyBac system facilitates the establishment of stable cell lines. A transgene flanked by piggyBac ITRs is randomly inserted into the host genome by a hyperactive piggyBac transposase (HyPBase), a more active form of PBase engineered by random mutagenesis [14]. The donor plasmid, pPB-TRE-λN-GOI, is co-transfected into the OSC line carrying the reporter gene, along with the helper plasmid (pAcM-HyPBase), to generate a Dox-inducible cell line for monitoring co-transcriptional silencing (Fig. 4b). The selection of cell lines using puromycin is essentially the same as the selection with blasticidin described in Subheading 3.3, with the added presence of a low concentration of blasticidin (5 μg/mL) for maintaining the boxB-luciferase reporter gene in the OSC genome during puromycin selection.

1. Plate approximately 1.0×10^6 OSCs carrying the reporter gene on a ϕ35 mm culture dish and culture at 26 °C overnight in the presence of 5 μg/mL blasticidin.
2. Confirm that the cells are approximately 60% confluent at the time of transfection.
3. Mix the following reagents in tubes.

Tube 1

Xfect reaction buffer	75 μL
pPB-TRE-λN-GOI (1 μg/μL)	1.5 μL
pAcM-HyPB (1 μg/μL)	1.5 μL

Tube 2

Xfect reaction buffer	75 μL
Xfect polymer	0.9 μL

4. Mix thoroughly by vortexing for 10 s at high speed.
5. Transfer the solution from Tube 2 to Tube 1.
6. Mix thoroughly by vortexing for 10 s at high speed.
7. Incubate for 10 min at room temperature to allow nanoparticle complexes to form.
8. Remove the OSC culture medium by aspiration, wash the cells twice with 2 mL PBS, and replace the medium with 1.5 mL of Shields and Sang M3 Insect Medium, to remove serum from the cell culture (*see* **Note 5**).
9. Add the entire nanoparticle complex solution dropwise to the cell culture medium. Rock the plate gently back and forth to mix.
10. Incubate OSCs at 26 °C for 3 h.

11. Remove the nanoparticle-containing medium by aspiration and replace the medium with 2 mL of fresh complete OSC culture medium containing 5 μg/mL blasticidin. Incubate OSCs at 26 °C for 48 h.
12. Two days after transfection, passage 25% of the cells in 8 mL of OSC culture medium containing 5 μg/mL puromycin and 5 μg/mL blasticidin to a ϕ100 mm culture dish.
13. Culture OSCs for 48 h at 26 °C.
14. Replate 25% of the cells in 8 mL of OSC culture medium containing 5 μg/mL puromycin and 5 μg/mL blasticidin on a ϕ100 mm culture dish.
15. Culture OSCs for 48 h at 26 °C.
16. Plate 2.0×10^6 OSCs in 8 mL of OSC culture medium containing 5 μg/mL puromycin and 5 μg/mL blasticidin on a ϕ100 mm culture dish (*see* **Note 9**).
17. Culture OSCs for 48 h at 26 °C.
18. Change the culture medium with fresh OSC medium containing 1 μg/mL puromycin and 5 μg/mL blasticidin.
19. Culture OSCs for at least 5 days until colonies reach a size that is easy to pick (Fig. 3b).
20. Under the microscope, scrape a well-isolated colony with a pipette tip onto a P20 pipette and transfer the cells to a tube containing 30 μL of 0.05% (w/v) Trypsin-EDTA solution.
21. Incubate the tube at 37 °C for 2 min and transfer the cell suspension to a well of a 24-well plate with 500 μL OSC culture medium containing 1 μg/mL puromycin and 5 μg/mL blasticidin.
22. Culture OSCs for 3–4 days. Replace the medium with fresh medium containing 1 μg/mL puromycin and 5 μg/mL blasticidin every 2 days.
23. Select cell lines with luciferase activity using the method described in Subheading 3.6.
24. Screen monoclonal OSC lines for expression of the transgene upon 1 μg/mL Dox induction for 24 h as described in Subheading 3.6 (*see* also Fig. 5c).
25. Expand positive clones. OSCs can be stored in liquid nitrogen using Cellbanker 1 cryopreservation medium, following the manufacturer's instructions.

3.6 Dox-Inducible Co-transcriptional Repression

Here, we describe a method to monitor co-transcriptional silencing using a cell line carrying a 14boxB-luciferase reporter and a λN-HA-3Flag-Panx gene driven by the TRE promoter (Fig. 5a). Panx is known to be tethered to transposable elements by the Piwi-piRNA complex, resulting in transcriptional silencing [3, 4]. In

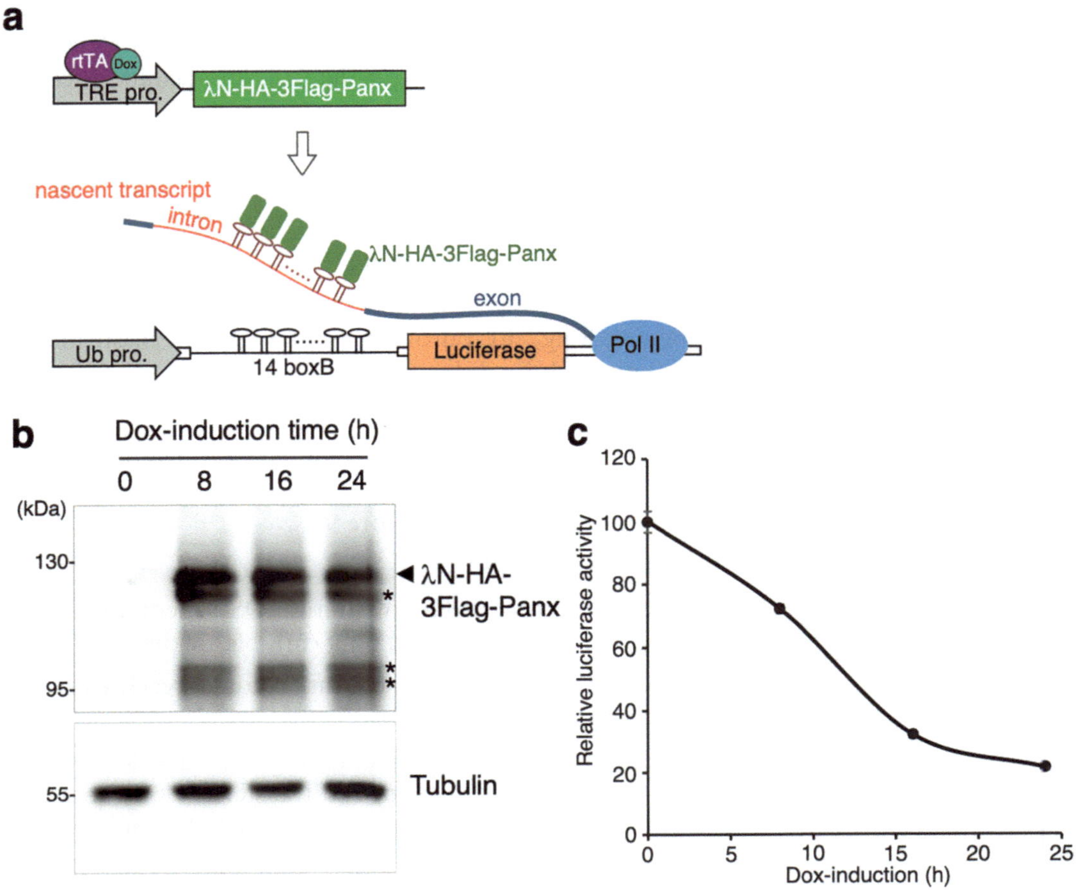

Fig. 5 Inducible expression of λN-HA-3Flag-Panx results in silencing of the intronic boxB reporter. **a** Schematic representation of the Dox-inducible λN-boxB tethering system. We established a tethering system in OSCs carrying a genome-integrated luciferase gene, which contains 14 intronic boxB sites (*see* **Note 3**). Expression of λN-HA-3Flag-Panx is induced by Dox. The λN fusion proteins are tethered to the nascent transcript of the reporter gene through the interaction of λN with boxB RNA. **b** Western blotting of λN-HA-3Flag-Panx with an anti-Flag antibody. Expression of λN-HA-3Flag-Panx was induced by the addition of 1 μg/mL Dox. Asterisks highlight degradation products originating from λN-HA-3Flag-Panx. Overexpressed λN-HA-3Flag-Panx is highly prone to degradation. β-Tubulin (Tub) was used as a loading control. **c** Luciferase activity gradually decreased after Dox treatment

brief, expression of λN-HA-3Flag-Panx is induced by the addition of 1 μg/mL Dox, and luciferase activity is used as a readout of repression. Forced tethering of Panx via boxB-λN causes repression of the reporter gene, which is easily identified as a drop in luciferase activity (Fig. 5b and c). Thus, this reporter system mimics transposon silencing by Panx. In addition to its application in OSCs, as demonstrated here, this method can also be applied to analyze in vivo effects within *Drosophila* ovaries (*see* **Note 10**).

1. Seed 2×10^5 OSCs on a 24-well plate with 500 μL of OSC culture medium.

2. On the next day, add Dox at a concentration of 1 μg/mL.
3. At the desired Dox-induction time, remove the culture medium and wash the cell layer with PBS once. Directly lyse the cells by adding 150 μL of Glo Lysis Buffer.
4. Incubate for 10 min at room temperature with vigorous shaking at 120 rpm on a plate shaker.
5. Transfer the lysate to a 1.5-mL tube. Centrifuge at 15 krpm for 5 min at 4 °C and transfer the supernatant to a new 1.5-mL tube.
6. Add 15 μL of the ONE-Glo EX Luciferase Assay reagent to a well of a 96-Well Polypropylene Microplate, White, Round bottom (*see* **Note 11**).
7. Add 15 μL of the supernatant to the well. Mix by pipetting 5 times.
8. Place the 96-well plate in the luminometer (Cytation 5, Bio-Tek) and initiate reading.
9. To compare luciferase activity between samples, normalize the activity by protein quantity. The protein quantity is measured using the Bradford method with the Protein Assay Dye Reagent Concentrate and the iMark microplate reader (Bio-Rad).

4 Notes

1. Sequences of plasmids can be downloaded as SnapGene files from GitHub (https://github.com/ch025572/2024_Tethering_method), and the plasmids can be distributed upon request. For construction of the piggyBac donor plasmid (pPB-TRE-lamN-HA-3Flag-EcoRV-Tjen-3G-P2A-puroR), OSC_Reporter_UAS_trafic jam_BoxB_d2eGFP_t2a_Blast (Addgene plasmid #128010) [5] and PB-P(tetO)-hDLX2-pA-floxPG-Kneo-pA [13] were used. For the construction of the helper plasmid, pCMV-HyPBase-PGK-Puro [13] was used.
2. Any high-fidelity PCR enzyme, such as PrimeSTAR Max DNA Polymerase (TAKARA, R045A), can be used. Perform the PCR reaction according to the manufacturer's instructions. We amplified a 5× boxB fragment by PCR using fly genomic DNA as a template (Vienna Drosophila Resource Center #313408: If/CyO; tub>EGFP_5xBoxB_SV40 [attP2]/TM3, Ser). Plasmids encoding 5× boxB sites are available from Addgene (ex. BoxB in pcDNA3, Addgene #29727 [15]).
3. Although insertions of multiples of five copies were expected, vector construction through PCR amplification of the 5× boxB

resulted in unexpected products. When amplifying DNA fragments containing repeat sequences such as the 5× boxB by PCR, it can happen that the number of repeats does not match the expected count. After construction of the vector, it is therefore necessary to confirm the copy number of the inserted fragment using Sanger sequencing. In our case, sequencing of the reporter plasmid obtained by repeating this insertion three times revealed that 14 copies of boxB had been inserted, instead of the expected 15. We are currently proceeding with analysis using this reporter plasmid, pUb-14boxB-Luc (Figs. 1b and 5a, *see* **Note 1**).

4. According to the manufacturer's instructions for Ligation High (TOYOBO), mix five times the amount of insert DNA relative to vector DNA in terms of molar ratio.
5. Although it is not recommended to remove serum from the culture medium in the manufacturer's protocol for the Xfect transfection reagent, removal increases the efficiency of plasmid transfection in the case of OSCs.
6. Cell death begins gradually, and the presence of blasticidin is exclusively tolerated by cells that have integrated the blasticidin-resistance gene into their genome.
7. When utilizing the NEBuilder HiFi DNA Assembly Master Mix for cloning, design primers to amplify the gene of interest according to the specifications indicated below.

   Primer F: GACAAGCTTGGTACCGATXXXX.
   Primer R: ATCATGTCTGGATCCGATXXXX.

These primers encompass the region surrounding the EcoRV site on the piggyBac donor plasmid (Fig. 4a), facilitating vector assembly by insertion. "XXXX" of the primer sequence denotes 18–25 bp sequences corresponding to both the beginning and the end of the coding sequence of the gene of interest. Refer to the manufacturer's protocol (https://www.neb.com/en/products/e2621-nebuilder-hifi-dna-assembly-master-mix) for comprehensive guidance on primer design.

8. DNA molar concentration can be calculated using the following equation: (fmol DNA) = (ng DNA) × 10^6 / 660 / Length (bp). For example, 6 ng of a 2000 bp DNA fragment is equivalent to approximately 5 fmol.
9. Cell death begins gradually, and the presence of puromycin is exclusively tolerated by cells that have integrated the puromycin-resistance gene into their genome. A reduced concentration of blasticidin (5 μg/mL) is employed for maintaining the boxB-luciferase reporter gene in the OSC genome during puromycin selection.

10. This tethering reporter system can also be used to analyze in vivo effects within *Drosophila* ovaries, as demonstrated in previous studies [3–8, 11, 16]. Using this system in germline cells of the fly ovary, a GFP sensor is significantly repressed when λN-fused Panx is expressed.
11. When using a transparent 96-well plate for the luciferase assay, light tends to leak into adjacent wells, causing interference between wells. To mitigate this interference, it is recommended to use a white, opaque 96-well plate.

Acknowledgments

We thank Haruhiko Siomi and Therese Solberg for their critical reading of the manuscript. YWI is supported by funding from JSPS KAKENHI Grant Number 22H02547, 25H01305, and JST Grant Numbers JPMJPR20E2 and JPMJFR224L. KM is supported by funding from JSPS KAKENHI Grant Number 23H02670.

References

1. Iwasaki YW, Siomi MC, Siomi H (2015) PIWI-interacting RNA: its biogenesis and functions. Annu Rev Biochem 84:405–433. https://doi.org/10.1146/annurev-biochem-060614-034258
2. Ozata DM et al (2019) PIWI-interacting-RNAs: small RNAs with big functions. Nat Rev Genet 20(2):89–108. https://doi.org/10.1038/s41576-018-0073-3
3. Sienski G et al (2015) Silencio/CG9754 connects the Piwi-piRNA complex to the cellular heterochromatin machinery. Genes Dev 29(21):2258–2271. https://doi.org/10.1101/gad.271908.115
4. Yu Y et al (2015) Panoramix enforces piRNA-dependent cotranscriptional silencing. Science 350(6258):339–342. https://doi.org/10.1126/science.aab0700
5. Batki J et al (2019) The nascent RNA binding complex SFiNX licenses piRNA-guided heterochromatin formation. Nat Struct Mol Biol 26(8):720–731. https://doi.org/10.1038/s41594-019-0270-6
6. Fabry MH et al (2019) piRNA-guided co-transcriptional silencing coopts nuclear export factors. elife 8. https://doi.org/10.7554/eLife.47999
7. Murano K et al (2019) Nuclear RNA export factor variant initiates piRNA-guided co-transcriptional silencing. EMBO J 38(17): e102870. https://doi.org/10.15252/embj.2019102870
8. Zhao K et al (2019) A pandas complex adapted for piRNA-guided transcriptional silencing and heterochromatin formation. Nat Cell Biol 21(10):1261–1272. https://doi.org/10.1038/s41556-019-0396-0
9. Iwasaki YW et al (2021) Piwi-piRNA complexes induce stepwise changes in nuclear architecture at target loci. EMBO J 40(18): e108345. https://doi.org/10.15252/embj.2021108345
10. Onishi R et al (2020) Piwi suppresses transcription of Brahma-dependent transposons via Maelstrom in ovarian somatic cells. Sci Adv 6(50). https://doi.org/10.1126/sciadv.aaz7420
11. Schnabl J et al (2021) Molecular principles of Piwi-mediated cotranscriptional silencing through the dimeric SFiNX complex. Genes Dev 35(5–6):392–409. https://doi.org/10.1101/gad.347989.120
12. Baron-Benhamou J, Gehring NH, Kulozik AE, Hentze MW (2004) Using the lambdaN peptide to tether proteins to RNAs. Methods Mol Biol 257:135–154. https://doi.org/10.1385/1-59259-750-5:135
13. Takeuchi C et al (2022) Generation of stable drosophila ovarian somatic cell lines using the piggyBac system. Methods Mol Biol 2509:

143–153. https://doi.org/10.1007/978-1-0716-2380-0_9

14. Yusa K et al (2011) A hyperactive piggyBac transposase for mammalian applications. Proc Natl Acad Sci USA 108(4):1531–1536. https://doi.org/10.1073/pnas.1008322108
15. Wang KC et al (2011) A long noncoding RNA maintains active chromatin to coordinate homeotic gene expression. Nature 472(7341):120–124. https://doi.org/10.1038/nature09819
16. Saito K (2022) Drosophila genetic resources for elucidating piRNA pathway. Methods Mol Biol 2509:135–141. https://doi.org/10.1007/978-1-0716-2380-0_8

Chapter 14

Machine Learning Approaches for Predicting RNA–RNA/DNA Interactions

Tsukasa Fukunaga

Abstract

In this chapter, I introduce machine learning approaches for predicting RNA–RNA/DNA interactions, which are crucial for understanding noncoding RNA (ncRNA) functions. Advancements in deep learning techniques and the availability of large-scale interaction data from high-throughput sequencing methods have driven the development of these prediction tools. This review covers representative studies across different RNA families, including prokaryotic small RNAs (TargetRNA3), general RNA–RNA interactions (CheRRI), miRNAs (DeepMirTar), box C/D snoRNAs (snoGloBe), lncRNA–DNA triplexes (triplexFPP), and CRISPR guide RNA design (CRISOT). These machine learning-based methods often improve accuracy compared to traditional energy-based approaches. However, there are challenges such as the need for preventing overfitting and third-party validation. Future advancements are expected to enhance the generalization and applicability of these prediction tools, contributing to a deeper understanding of RNA functions.

Key words RNA–RNA interaction, RNA triplex, Noncoding RNA, Machine learning, Deep learning

1 Introduction

Many ncRNAs function by interacting with other biomolecules. Among these interactions, those with other RNA/DNA through complementary base-pairing often play an essential role due to their high target specificity. For example, miRNAs function to repress gene expression by binding to the 3′-UTRs of target mRNAs [1]. In addition, RNA–DNA triple helices (triplexes) have functions such as recruiting transcription factors to regulate gene expression and targeting epigenetic modifiers [2]. Furthermore, the genome editing tool CRISPR-Cas9 induces DNA double-strand breaks based on base-pairing between target DNA regions and designed sgRNAs [3]. To understand the functions of these ncRNAs and to design efficient sgRNAs, various bioinformatics tools have been developed to predict these interactions [4–

Michiaki Hamada et al. (eds.), *RNA-RNA and RNA-DNA Interactions*, Methods in Molecular Biology, vol. 2949,
https://doi.org/10.1007/978-1-0716-4670-0_14,

7]. Many of these methods are based on physical energy models, such as the neighborhood energy model [8], or rule-based models.

The RNA secondary structure prediction problem, which predicts base-pairing within RNA molecules, is similar to the prediction of RNA–RNA/DNA interactions. In this problem, as with interaction prediction, many programs use nearest neighbor energy models and predict structures based on the principle of minimum free energy or maximum expected accuracy [9, 10]. On the other hand, research on secondary structure prediction based on machine learning has also been actively conducted, and in recent years, the development of prediction methods utilizing deep learning has become a hot topic [11, 12]. For example, MXfold2 achieved robust and accurate secondary structure prediction by integrating energy models and deep learning [12].

Although it is not as actively studied as RNA secondary structure prediction, RNA–RNA/DNA interaction prediction based on machine learning has recently attracted attention. This is largely due not only to advances in deep learning techniques but also to the availability of large-scale RNA interaction data through high-throughput sequencing methods. For example, the PARIS method enables the identification of interacting RNAs in living cells by proximity ligation [13]. In addition, the Triplex-seq method identifies RNAs that form triplexes by excluding free RNAs, extracting chromatin-associated RNAs, and then excluding R-loop-associated RNAs [14]. In this brief review, we introduce representative studies that predict RNA–RNA/DNA interactions using machine learning approaches and discuss the potential for future developments.

2 Representative Machine Learning Approaches for Predicting RNA–RNA/DNA Interactions

2.1 TargetRNA3

Small RNAs (sRNAs) are widely found in prokaryotes, and in *E. coli*, hundreds of sRNAs are involved in posttranscriptional regulation through RNA–RNA interactions. The advancement of RNA-seq technology has facilitated the detection of sRNAs in diverse organisms. However, the targets of most sRNAs remain unknown, although detection techniques for their interactions, such as RIL-seq using co-IP against the sRNA-binding protein Hfq, have been developed [15]. Therefore, computational interaction prediction is important and has been studied for many years.

TargetRNA is an energy model-based interaction prediction tool that has been widely used since the first paper was published in 2006 [16]. The developers have recently released the updated version, TargetRNA3, which incorporates machine learning into RNA-RNA interaction prediction [17]. TargetRNA3 predicts interactions using 111 features, including hybridization energies, 3-mer occurrence frequencies, and distances to the start codon, through a gradient boosting method. The dataset consists of 4386 positive interactions and 325,162 negative interactions from 77 sRNAs across 13 genomes. The authors showed that TargetRNA3 outperforms existing tools in accuracy and is nearly 50 times faster than current high-accuracy tools. In addition, TargetRNA3 is easily usable by experimental biologists through its web service (https://cs.wellesley.edu/~btjaden/TargetRNA3).

On the other hand, as the authors discussed in the paper, TargetRNA3 has some limitations. First, the number of genomes used for training was 13, which was very small relative to the diversity of prokaryotes. Whether the prediction results can be extrapolated to sRNAs that are independent on RNA-binding proteins, such as Hfq, or to sRNAs in archaea for which there is little experimental data on interactions, should be verified in the future. In addition, although TargetRNA3 is more accurate than other methods, it is still not accurate enough, and further improvement is needed.

2.2 CheRRI

As previously mentioned, many RNA–RNA interaction prediction tools have been developed, but benchmarking studies have shown that their prediction is not sufficiently reliable [18]. In particular, the large number of false positives, where regions are predicted to interact stably but do not actually interact, poses a significant challenge for experimental work. This is likely due to various biological phenomena that are difficult to account for with energy models but indeed influence RNA–RNA interactions. For example, there are the co-transcriptional folding process, RNA tertiary structures, interactions with chromatin or RNA binding proteins, and RNA modifications.

CheRRI is a machine learning method to estimate whether the interaction sites predicted by IntaRNA [4], the standard prediction tool, are reliable [19]. The interaction dataset consists of human HEK293 T cell and mouse embryonic stem cell data obtained by the PARIS method, each including tens of thousands of positive and negative data [13]. In addition, CheRRI can incorporate RNA-binding protein binding region data into the prediction model. The features used for learning include those derived from the secondary structure, such as hybridization energy and accessibility energy, and those derived from the primary sequence, such as GC content and GC skew. Additionally, CheRRI can use subgraph feature kernels, which is calculated by regarding the predicted secondary structure as a graph, for the features. CheRRI can be installed via pip or conda and is also easily accessible through Galaxy, the web-based bioinformatics analysis platform [20].

The authors demonstrated that the F1 scores for CheRRI prediction exceeded 0.9 when the training and test data were derived from the same species. They also showed that the prediction accuracy was slightly better when a subgraph feature kernel was used. On the other hand, when the training and test data were from different species, the F1 scores dropped to about 0.6. Therefore, the generalization performance of CheRRI on diverse datasets requires further investigation. Specifically, it is important to examine its predictive performance on data from different cell types in the same species and its applicability to prokaryotic RNA–RNA interaction prediction. Additionally, evaluating the predictive performance for functional RNA–RNA interaction and assessing how the performance changes when using interaction detection methods other than PARIS are essential validation issues.

2.3 DeepMirTar

MiRNAs are 23 nt ncRNAs that bind to the 3′-UTR of target genes and repress their translation [21, 22]. MiRNAs are one of the best-understood ncRNAs, with many studies having been performed on their biosynthetic processes and mechanisms of action. In the bioinformatics field, numerous miRNA target prediction software tools have also been developed, incorporating experimentally obtained biological knowledge as features. For example, the complementarity of the 7-nucleotide region called the seed region is a frequently used feature. Additionally, the target regions of miRNAs are evolutionarily well conserved, so sequence conservation is also used as a feature. The application of deep learning to miRNA target prediction research was published early in Ref. [23], but initial studies were noted to have potential biases in the methods of constructing datasets.

DeepMirTar is an miRNA target prediction tool published in 2018 that uses a stacked de-noising auto-encoder as the learning model [24]. For the positive data, they used 3915 interactions from known low-throughput experiment-based and high-throughput sequencing-based data. For the negative data, mock miRNAs were generated, and 3905 negative data were created by searching for regions that satisfy the seed match condition. A total of 750 features were used, including RNA accessibility, seed match, free energy, conservation, and one-hot encoding. As a result, the authors showed that the accuracy of DeepMirTar was higher compared to other learning algorithms, such as Random Forest and CNN, and existing miRNA-target prediction tools. In addition, feature analysis using Random Forest revealed that not only the traditionally important seed region but also one-hot encoding significantly contributed to improving accuracy.

Compared to the 2000s when miRNAs were the focus of much attention, the development of miRNA target prediction tools has not received as much attention in recent years, and there are not many examples of studies using deep learning. However, this does not mean that these prediction tools are not used by experimental biologists. For example, the standard miRNA-interaction prediction tool TargetScan [25] was cited 761 times in Google Scholar in 2023 alone. One reason for its widespread use is that TargetScan has a user-friendly web service. For deep learning-based tools to be widely adopted, it is essential to prepare a user-friendly environment.

2.4 snoGlobe

SnoRNAs induce methylation or pseudouridylation of RNA genes such as rRNA and snRNA [26], and they identify the modification sites by interacting with target RNAs. Although numerous snoRNAs have been discovered with the advancement of small ncRNA-seq, many of them have not yet found their targets and are called orphan snoRNAs [27]. SnoRNAs are classified into box C/D snoRNAs and box H/ACA snoRNAs based on their secondary structures and the type of RNA modification. In box C/D snoRNAs, it has been shown that a region of about ten nucleotides between two sequence motifs (Box C and Box D) interacts with target RNAs. Recently, however, noncanonical interactions involving various regions, such as those spanning the box C/D motifs, have been discovered. These interactions play diverse roles, such as binding to mRNA, to regulate alternative splicing [28]. Nevertheless, the development of tools to predict snoRNA interactions has been very limited, especially for predicting noncanonical interactions.

snoGloBe is the first bioinformatics tool that can predict both noncanonical and canonical interactions [29]. For positive data, snoGloBe used interactions obtained from high-throughput sequencing methods such as PARIS [13]. Because these interaction data contain a lot of noise, sequences that do not form complementary base pairs were excluded using RNAplex [30]. They also included known interactions obtained from small-scale experiments to the dataset, resulting in a total of 594 interaction data. Negative data were created at a ratio of 1000 negative data for each positive data, considering that snoRNAs do not bind to most RNAs. As features, snoGloBe used one-hot encoding of the sequence, relative position within the snoRNA, target biotype, and target position (e.g., CDS or 5′ UTR), and used gradient boosting as the learning model.

snoGloBe showed better accuracy compared to other tools that can only predict canonical interactions and general RNA–RNA interaction prediction tools. When the authors applied snoGloBe to all genes for the highly expressed snoRNAs, an average of 1017 interactions per snoRNA were predicted, and one snoRNA had over 10,000 predicted interactions. The interactions were enriched in the 5′-UTR and exon-intron junctions, suggesting a potential role in translation/transcription regulation and alternative splicing. In addition, GO term analysis suggested that certain snoRNAs are associated with specific functions; for example, target genes of SNORD50B were enriched in neuronal functionality and cell–cell interaction. Moreover, changes in gene expression levels upon

SNORD126 knockdown were shown to be associated with the predicted interactions.

Although snoRNAs have been known for a long time and some of them show high expression levels, many of their functions remain unknown. The use of snoGloBe is expected to further elucidate the functions of snoRNAs. In addition, developing more advanced prediction tools for box H/ACA snoRNAs is important.

2.5 triplexFPP

Long noncoding RNAs (lncRNAs) are ncRNAs longer than 500 bases, and there are tens of thousands of lncRNAs in humans [31]. Some lncRNAs have known functions, such as XIST, which is involved in X-chromosome inactivation, and Neat1, which is associated with the formation of nuclear paraspeckles, but about 99% of lncRNAs have no understood function at all [32]. Bioinformatics is a powerful approach to infer lncRNA function; however, due to the low sequence conservation of lncRNAs, sequence similarity analysis is not effective for lncRNA function analysis. Therefore, prediction of RNA and DNA interactions with the lncRNA is widely used. The binding modes between RNA and DNA include the formation of R-loops, where RNA and DNA form a double strand with a single-stranded DNA region, and the formation of triplex structures based on the Hoogsteen base-pairing rule. Fendrr is an lncRNA that forms a triplex structure, and this triplex serves as an epigenetic mark involved in embryogenesis [33].

TriplexFPP is a machine learning method for predicting lncRNA and DNA regions that form triplex structures [34]. This method uses *k*-mer-based scores as features and employs a two-layer convolutional neural network as the learning model. For the training data, both large-scale data obtained from Triplex-seq [14] and small-scale experimentally validated data were used. The dataset was curated by excluding sequences without DNA binding domains using the Triplex domain finder [35] and clustering similar sequences to remove redundancy. Finally, they constructed a dataset consisting of hundreds of positive data and tens of thousands of negative data.

The authors showed that TriplexFPP had high prediction accuracy for both DNA regions and lncRNAs in the input dataset. However, TriplexFPP has the limitation that it only determines if a region is likely to form the triplex but cannot predict how the triplex forms. In addition, the authors have not reported novel lncRNAs that are likely to form triplexes. Furthermore, the prediction accuracy is high, although the problem does not seem overly simple, suggesting a potential issue of overfitting that may need further investigation. Further research is required on learning models related to triplex formation and their applications.

2.6 CRISOT

The CRISPR-Cas9 system is a widely used genome editing tool that employs a guide RNA complementary to the target DNA sequence. In this system, designing the guide RNA is a critical issue because unintended genome editing can occur when the complementary binding occurs at off-target sites. The difficulty in this design problem is that guide RNAs can bind to target DNA sequences even if they are not completely complementary but have some mismatches. At present, various tools for designing sgRNAs, such as CRISPRoff [36] based on energy models and DeepCRISPR [37] based on deep learning, have been developed, but their accuracy has not been sufficient.

Chen et al. proposed CRISOT [38], which improves the prediction accuracy of genome editing by considering the molecular mechanisms within the Cas9 complex. An interesting point of CRISOT is the use of molecular dynamics (MD) simulation results as machine learning features. They performed MD simulations on the Cas9-sgRNA–DNA complex structures after varying the bases of the RNA–DNA hybrid and used the simulation results, such as atomic positions and binding free energies, as features. High-throughput sequencer-based methods such as Change-seq [39] were used to estimate Cas9-mediated DNA cleavages as the training data. They used Xgboost as the learning algorithm and showed that CRISOT predicts cleavages more accurately than existing methods.

They also proposed CRISOT-Spec, which estimates how likely a designed sgRNA is to cause off-target effects, and CRISOT-opti, which modifies the proposed sgRNA to reduce off-target effects. In addition, they demonstrated that CRISOT can estimate off-target effects, even for base editor and prime editor, which edit bases without cleaving DNAs.

One-hot vectors and *k*-mer counts are often used to convert nucleotide sequences into numerical features. Recently, representation learning such as RNABERT [40] has been increasingly used, but compared to the success in protein sequences, research in RNA sequences is still in its early stages. The use of MD simulation results as features is interesting research, and it is expected that this approach will be extended to other RNAs beyond CRISPR in the future.

3 Conclusion

In this review, we briefly introduced tools for predicting RNA–RNA/DNA interactions using machine learning methods. The development of these methods is due not only to progress in machine learning techniques but also to the accumulation of large-scale data based on advances in experimental techniques. The benchmarking showed that machine learning-based methods can predict with higher accuracy compared to the previously dominant energy-based and rule-based methods. On the other hand, since RNA binding characteristics differ among RNA families, the development of prediction methods specialized for each family is necessary, but this area is still insufficient and further progress is expected in the future.

When using machine learning, it is generally important to be careful to prevent overfitting. This happens when the learning process becomes overly adapted to the training data, resulting in poor generalization performance on unknown test data. In this context, overfitting means that predictions for RNA families or subfamilies included in the training data may be accurate, but predictions for other unknown RNA families or subfamilies are significantly less accurate. In RNA secondary structure prediction, many software programs based on machine learning are suspected of overfitting [41, 42]. To prevent overfitting in RNA–RNA/DNA interaction prediction, the learning process, especially the way the datasets for training and testing data are divided, must be designed with great care. In addition, third-user validation is needed to ensure that existing tools are not overfitting.

References

1. Gebert LFR, MacRae IJ (2019) Regulation of microRNA function in animals. Nat Rev Mol Cell Biol 20(1):21–37
2. Leisegang MS, Warwick T, Stötzel J, Brandes RP (2024) RNA-DNA triplexes: molecular mechanisms and functional relevance. Trends Biochem Sci 49(6):532–544
3. Doudna JA, Charpentier E (2014) The new frontier of genome engineering with CRISPR-Cas9. Science 346(6213):1258096
4. Mann M, Wright PR, Backofen R (2017) IntaRNA 2.0: enhanced and customizable prediction of RNA-RNA interactions. Nucleic Acids Res 45(W1):W435–W439
5. Fukunaga T, Hamada M (2017) RIblast: an ultrafast RNA-RNA interaction prediction system based on a seed-and-extension approach. Bioinformatics 33(17):2666–2674
6. Buske FA, Bauer DC, Mattick JS, Bailey TL (2012) Triplexator: detecting nucleic acid triple helices in genomic and transcriptomic data. Genome Res 22(7):1372–1381
7. Naito Y, Hino K, Bono H, Ui-Tei K (2015) CRISPRdirect: software for designing CRISPR/Cas guide RNA with reduced off-target sites. Bioinformatics 31(7): 1120–1123
8. Mittal A, Turner DH, Mathews DH (2024) NNDB: the nearest neighbor parameter database for predicting stability of nucleic acid secondary structure. J Mol Biol 38(Database issue):D280–D282
9. Zuker M (2003) Mfold web server for nucleic acid folding and hybridization prediction. Nucleic Acids Res 31(13):3406–3415
10. Hamada M, Kiryu H, Sato K, Mituyama T, Asai K (2009) Prediction of RNA secondary structure using generalized centroid estimators. Bioinformatics 25(4):465–473
11. Singh J, Hanson J, Paliwal K, Zhou Y (2019) RNA secondary structure prediction using an ensemble of two-dimensional deep neural networks and transfer learning. Nat Commun 10(1):5407
12. Sato K, Akiyama M, Sakakibara Y (2021) RNA secondary structure prediction using deep learning with thermodynamic integration. Nat Commun 12(1):941
13. Zhipeng L, Zhang QC, Lee B, Flynn RA et al (2016) RNA duplex map in living cells reveals higher-order transcriptome structure. Cell 165(5):1267–1279
14. Cetin NS, Kuo C-C, Ribarska T, Li R et al (2019) Isolation and genome-wide characterization of cellular DNA:RNA triplex structures. Nucleic Acids Res 47(5):2306–2321
15. Melamed S, Peer A, Faigenbaum-Romm R, Gatt YE et al (2016) Global mapping of small

RNA-target interactions in bacteria. Mol Cell 63(5):884–897
16. Tjaden B, Goodwin SS, Opdyke JA, Guillier M et al (2006) Target prediction for small, noncoding RNAs in bacteria. Nucleic Acids Res 34(9):2791–2802
17. Tjaden B (2023) TargetRNA3: predicting prokaryotic RNA regulatory targets with machine learning. Genome Biol 24(1):276
18. Lai D, Meyer IM (2016) A comprehensive comparison of general RNA-RNA interaction prediction methods. Nucleic Acids Res 44(7): e61
19. Müller T, Mautner S, Videm P, Eggenhofer F et al (2024) CheRRI-Accurate classification of the biological relevance of putative RNA-RNA interaction sites. GigaScience 13:giae022
20. Galaxy Community (2022) The Galaxy platform for accessible, reproducible and collaborative biomedical analyses: 2022 update. Nucleic Acids Res 50(W1):W345–W351
21. Kehl T, Backes C, Kern F, Fehlmann T et al (2018) About miRNAs, miRNA seeds, target genes and target pathways. Oncotarget 8(63): 107167–107175
22. Shang R, Lee S, Senavirathne G, Lai EC (2023) microRNAs in action: biogenesis, function and regulation. Nat Rev Genet 24(12):816–833
23. Cheng S, Guo M, Wang C, Liu X et al (2016) MiRTDL: a deep learning approach for miRNA target prediction. IEEE/ACM Trans Comput Biol Bioinform 13(6):1161–1169
24. Wen M, Cong P, Zhang Z, Hongmei L, Li T (2018) DeepMirTar: a deep-learning approach for predicting human miRNA targets. Bioinformatics 34(22):3781–3787
25. Agarwal V, Bell GW, Nam J-W, Bartel DP (2015) Predicting effective microRNA target sites in mammalian mRNAs. elife 4:e05005
26. Maxwell ES, Fournier MJ (1995) The small nucleolar RNAs. Annu Rev Biochem 64:897–934
27. Bachellerie JP, Cavaillé J, Hüttenhofer A (2002) The expanding snoRNA world. Biochimie 84(8):775–790
28. Bratkovič T, Božič J, Rogelj B (2020) Functional diversity of small nucleolar RNAs. Nucleic Acids Res 48(4):1627–1651
29. Deschamps-Francoeur G, Couture S, Abou-Elela S, Scott MS (2022) The snoGloBe interaction predictor reveals a broad spectrum of C/D snoRNA RNA targets. Nucleic Acids Res 50(11):6067–6083
30. Tafer H, Hofacker IL (2008) RNAplex: a fast tool for RNA-RNA interaction search. Bioinformatics 24(22):2657–2663
31. Mattick JS, Amaral PP, Carninci P, Carpenter S et al (2023) Long non-coding RNAs: definitions, functions, challenges and recommendations. Nat Rev Mol Cell Biol 24(6):430–447
32. de Hoon M, Shin JW, Carninci P (2015) Paradigm shifts in genomics through the FANTOM projects. Mamm Genome 26(9–10): 391–402
33. Grote P, Herrmann BG (2013) The long non-coding RNA Fendrr links epigenetic control mechanisms to gene regulatory networks in mammalian embryogenesis. RNA Biol 10(10):1579–1585
34. Zhang Y, Long Y, Kwoh CK (2020) Deep learning based DNA:RNA triplex forming potential prediction. BMC Bioinformatics 21(1):522
35. Kuo C-C, Hänzelmann S, Cetin NS, Frank S et al (2019) Detection of RNA-DNA binding sites in long noncoding RNAs. Nucleic Acids Res 47(6):e32
36. Alkan F, Wenzel A, Anthon C, Havgaard JH, Gorodkin J (2018) CRISPR-Cas9 off-targeting assessment with nucleic acid duplex energy parameters. Genom Biol 19(1):177
37. Chuai G, Ma H, Yan J, Chen M et al (2018) DeepCRISPR: optimized CRISPR guide RNA design by deep learning. Genom Biol 19(1):80
38. Chen Q, Chuai G, Zhang H, Tang J et al (2023) Genome-wide CRISPR off-target prediction and optimization using RNA-DNA interaction fingerprints. Nat Commun 14(1): 7521
39. Lazzarotto CR, Malinin NL, Li Y, Zhang R et al (2020) CHANGE-seq reveals genetic and epigenetic effects on CRISPR-Cas9 genome-wide activity. Nat Biotechnol 38(11): 1317–1327
40. Akiyama M, Sakakibara Y (2022) Informative RNA base embedding for RNA structural alignment and clustering by deep representation learning. NAR Genom Bioinform. 4(1): lqac012 4
41. Rivas E, Lang R, Eddy SR (2012) A range of complex probabilistic models for RNA secondary structure prediction that includes the nearest-neighbor model and more. RNA 18(2):193–212
42. Szikszai M, Wise M, Datta A, War M, Mathews DH (2022) Deep learning models for RNA secondary structure prediction (probably) do not generalize across families. Bioinformatics 38(16):3892–3899

Chapter 15

Identification and Sequence Characterization of Semi-extractable RNAs

Chao Zeng and Michiaki Hamada

Abstract

We present customized bioinformatic protocols for the identification and sequence characterization of semi-extractable RNAs, which are promising in understanding RNA-centric phase separation. Through a comprehensive bioinformatic analysis of multiple public datasets, we describe the unique sequence characteristics and interaction profiles of these RNAs, shedding light on their biological significance. Our findings provide a foundation for future investigations into the molecular mechanisms of RNA-mediated phase separations and their implications in RNA biology. The associated analysis code can be accessed at (https://github.com/zengXchao/serna_mimb).

Key words Semi-extractable RNA, Phase separation, NEAT1

1 Introduction

The exploration of semi-extractable RNAs, a class of biomolecules increasingly recognized for their role in the biomolecular condensates (also called membrane-less organelles) through phase separation [1], requires sophisticated analytical approaches [2]. Such RNAs often remain partially sequestered within complex biomolecular assemblies, avoiding conventional detection methods and requiring advanced bioinformatic analyses to unravel their sequences and functional roles. To address these challenges, this chapter introduces a series of bioinformatic protocols designed to systematically identify and characterize semi-extractable RNAs from extensive transcriptomic datasets. The primary objective of these protocols is to enhance the resolution at which these RNAs can be investigated, thereby providing deeper insights into their biological functions and the mechanisms by which they contribute to the formation and maintenance of biomolecular condensates.

Our bioinformatic methodology integrates multiple layers of computational analysis, beginning with the detection of

Michiaki Hamada et al. (eds.), *RNA-RNA and RNA-DNA Interactions*, Methods in Molecular Biology, vol. 2949, https://doi.org/10.1007/978-1-0716-4670-0_15,

semi-extractable RNAs from RNA-seq data. We utilize publicly available experimental data and predictive tools to analyze their chromatin state distributions, subcellular localizations, and biomolecular interactions, including RNA–DNA and RNA–RNA interactions. Additionally, we evaluate the intrinsic features of the sequences such as length, minimal free energy (MFE), GC and k-mer content, and the presence of RNA-binding protein (RBP) motifs. Based on the k-mer content, we perform clustering analysis to categorize semi-extractable RNAs into distinct groups, potentially implying varied biological functions in the formation of biomolecular condensates. In conclusion, the bioinformatic analyses conducted in this chapter represent a significant advancement in our ability to decode the complex roles of semi-extractable RNAs.

2 Materials

2.1 Hardware

Linux operating system with multicore processors. It is recommended to use a machine with more than 64 GB of physical memory.

2.2 Software

1. Mamba, Python, and Pip. Mamba (https://github.com/mamba-org/mamba) are used to install and manage all the software packages used for the analysis. We utilize a lightweight Conda environment with Miniforge (https://github.com/conda-forge/miniforge) preinstalled with Mamba, Python, and Pip (https://github.com/pypa/pip). During the installation of the Miniforge environment, the default installation path can be set to *~/miniforge3* as prompted. This will ensure that all software is installed in the *~/miniforge3/bin* directory.

```
$ curl -L -O "https://github.com/conda-forge/miniforge/re-
leases/latest/download/Miniforge3-$(uname)-$(uname -m).sh"
$ bash Miniforge3-$(uname)-$(uname -m).sh
$ chmod +x ~/miniforge3/bin/*
$ eval "$(~/miniforge3/bin/conda shell.bash hook)"
$ conda init
```

2. Pandas [3], Numpy [4], Scipy [5], Biopython [6], Gffutils (https://github.com/fls-bioinformatics-core/GFFUtils), Csvkit (https://github.com/wireservice/csvkit), Matplotlib [7], Seaborn [8], Matplotlib_venn (https://github.com/konstantint/matplotlib-venn), Seekr [9], Networkx (https://networkx.org/), Python-louvain (https://github.com/taynaud/python-louvain). These packages can be installed using Pip. We use the —*user* option to avoid installation failures due to lack of administrative privileges. Additionally, we use the —*no-cache-dir* option to force Pip to download packages from

remote sources, preventing installation failures caused by outdated, incomplete, or incompatible cached packages.

```
$ pip install pandas biopython gffutils numpy csvkit scipy
seaborn matplotlib matplotlib_venn seekr networkx python-
louvain --no-cache-dir
```

3. R and edgeR. EdgeR [10] needs to be installed from within the R (https://www.r-project.org) interactive environment. During the installation process, follow the prompts to configure various parameters.

```
$ mamba install r
$ chmod +x ~/miniforge3/lib/R/bin/*
$ R
$ if (!requireNamespace("BiocManager", quietly = TRUE))
 install.packages("BiocManager")
 BiocManager::install("edgeR")
$ q()
```

4. Samtools [11], STAR [12], Stringtie [13], Subread [14], Bedtools [15], Cutadapt [16], Gffread [17], Sra-tools (https://github.com/ncbi/sra-tools), Ucsc-liftover (https://genome-store.ucsc.edu), Viennarna [18], and Meme [19]. These packages can be installed using Mamba from the bioconda channel (https://bioconda.github.io).

```
$ mamba install -c bioconda samtools star stringtie subread
bedtools cutadapt gffread sra-tools ucsc-liftover viennarna
meme
$ chmod +x ~/miniforge3/bin/*
```

3 Methods

3.1 Data Preparation

This section provides an overview of all public data used in this analysis. The data downloading and preprocessing steps have been automated and executed in the script *./run.sh*.

1. Semi-extractable and conventional RNA-seq

RNA-seq data obtained from two biological replicates of HEK cell lines, using both conventional and improved RNA extraction methods, were utilized in this analysis. The corresponding accession codes are DRA009793 and DRA012807, which can be accessed from the DDBJ Sequence Read Archive (DRA) at https://www.ddbj.nig.ac.jp.

2. Genomic sequences, gene annotation, and rRNA sequences.

The human genomic sequence (hg38) and the gene annotation GTF file (comprehensive, all regions) were obtained from release 43 of the GENCODE project (https://www.gencodegenes.org/human/release_43.html). Ribosomal RNA sequences were manually downloaded and merged from the NCBI RefSeq (https://www.ncbi.nlm.nih.gov/refseq/) and Ensembl BioMart (https://www.ensembl.org/biomart/martview) databases (accessed in May 2023).

3. Chromatin States

Chromatin states were sourced from the ENCODE project, accessible at (http://hgdownload.cse.ucsc.edu/goldenpath/hg19/encodeDCC/wgEncodeAwgSegmentation/wgEncodeAwgSegmentationChromhmmHelas3.bed.gz). The states were determined using a ChromHMM [20] model trained on multiple chromatin datasets, including ChIP-seq data for several histone modifications. The chromatin state categories were revised and labeled as follows: (i) Active Promoter: Tss and TssF; (ii) Promoter Flanking: PromF; (iii) Inactive Promoter: PromP; (iv) Potential Strong Enhancer: Enh and EnhF; (v) Potential Weak Enhancer/DNase: EnhWF, EnhW, DNaseU, DNaseD; (vi) Distal CTCF/Candidate Insulator: CtrcfO and Ctcf; (vii) Transcription-related: Gen5', Elon, ElonW, Gen3', Pol2, H4K20; (viii) Low Activity Near Active States: Low; (ix) Polycomb Repressed: ReprD, Repr, and ReprW; and (x) Heterochromatin/Repetitive Elements/Copy Number Variation: Quies, Art.

4. RNA subcellular localization (APEX-Seq)

APEX-Seq [21] data based on direct proximity labeling of RNA in HEK 293 T cells from various subcellular locations were downloaded from the NCBI SRA (https://www.ncbi.nlm.nih.gov/sra). The datasets and their corresponding accession codes are as follows: Nucleus (SRR7367268, SRR7367269, SRR7367270, SRR7367271), Nucleolus (SRR7367272, SRR7367273, SRR7367274, SRR7367275), Nuclear lumina (SRR7367276, SRR7367277, SRR7367278, SRR8646402), Nuclear pore (SRR7367280, SRR7367281, SRR7367282, SRR7367283), Cytosol (SRR7367284, SRR7367285, SRR7367286, SRR7367287), ER membrane cytosol-facing (SRR7367288, SRR7367289, SRR7367290, SRR7367291), Outer mitochondrial membrane (SRR7367296, SRR7367297, SRR7367298, SRR7367299), Mitochondrial matrix (SRR7367300, SRR7367301, SRR7367303, SRR7367304), and ER lumen (SRR7367292, SRR7367293, SRR7367294, SRR7367295).

5. RNA–DNA interaction

The processed iMARGI (in situ Mapping of RNA–Genome Interactions) [22] data for HEK 293 T cells can be accessed from NCBI GEO (https://ncbi.nlm.nih.gov/geo) under the accession number GSM3478205. The data are stored in BEDPE format, recording the positions of the two interacting regions in the genome.

6. RNA–RNA interaction

The list of hubRNAs in HeLa cells, calculated from RIC-seq [23] data (RNAs interacting with multiple other RNAs), can be directly obtained from Supplementary Table 7 of the original paper (https://static-content.springer.com/esm/art%3A10.1038%2Fs41586-020-2249-1/MediaObjects/41586_2020_2249_MOESM10_ESM.xlsx). The processed PARIS [24] data for HEK 293 T cells can be downloaded from the RISE [25] database (http://rise.life.tsinghua.edu.cn/download-data/rise_human_transcriptome.txt.gz).

7. RBP-binding motif

The motif database used for predicting RBP-binding sites in RNA sequences was downloaded from the MEME website (https://meme-suite.org/meme/meme-software/Databases/motifs/motif_databases.12.24.tgz). We selected the Homo sapiens dataset from CISBP-RNA as the reference.

8. Repetitive sequence

Genomic repeat regions and classification information (hg38, lib20240131) were downloaded from RepeatMasker (https://www.repeatmasker.org).

3.2 Detection of Semi-extractable RNAs from RNA-Seq

After constructing the transcriptome sequences using RNA-seq data, we performed quantitative and differential expression analyses on these transcripts, ultimately identifying the set of semi-extractable RNAs (SE) and the control group of extractable RNAs (EX). The detailed processing pipeline can be found in Part B of the main script file *./run.sh* and the scripts it calls. The following sections will highlight some key steps.

The downloaded RNA-seq files were decompressed using *gzip*, and adapter sequences were removed with *cutadapt*, retaining reads longer than 20 nt to avoid potential issues from short reads causing multiple alignments. Reads mapping to rRNA sequences were removed to prevent biases in subsequent analyses due to highly expressed rRNAs. Next, we set the *STAR* parameter "*—outSAMmultNmax 1*" to allow reads to map to multiple genomic locations but randomly select one alignment.

The mapping results were used with *StringTie* to assemble the transcriptomes of each sample. We then used the *—merge* mode of *StringTie* to combine the assembled transcriptomes from all samples (*see* **Note 1**). Transcripts overlapping with annotated genes were removed from the set assembled by *StringTie*, and nearby transcripts were merged into transcript clusters. These clusters represent newly detected transcripts in intergenic regions. For annotated gene regions, we used *StringTie* in *-e* mode to quantify the expression levels (FPKM, fragments per kilobase of exon/intron per million mapped reads) of all exons and introns from the mapping results. The retention score (RS) for *i*-th intron was calculated as follows:

$$RS_i = \frac{2 * I_i}{E_i + E_{i+1}}, \tag{1}$$

where E and I represent the expression levels of the intron and exon, respectively. Introns with both RS and FPKM values greater than 0.1 are considered to be expressed in the corresponding sample, indicating an intron retention event. These introns are incorporated into the transcript. The resulting transcripts that include introns are labeled with the suffix "_RI" in their transcript_id to distinguish them (*see* **Note 2**). Finally, the previously mentioned transcript clusters were combined with the annotated transcripts to form a reference transcriptome for further analysis.

Based on the above reference transcriptome and mapping data, we used *featureCounts* to quantify transcript read counts and then performed differential expression analysis with *edgeR*, resulting in the outcomes shown in Fig. 1. Transcripts with a log2 fold change greater than 0 and an FDR less than 0.05 were considered upregulated in the improved RNA-seq samples (labeled as Up, red) and termed semi-extractable RNAs. We then selected an equal number of nonsignificant transcripts (labeled as NS, blue) from the remaining transcripts, sorted in descending order by the absolute value of log2 fold change. Finally, we removed redundant transcripts (retaining the version with intron retention if both are present) from the Up and NS transcripts, resulting in the identification of semi-extractable RNAs (SE) and extractable RNAs (EX).

3.3 Chromatin State Analysis of Semi-extractable RNAs

To determine the overlap between chromatin states and RNA transcription regions, we first need to convert the chromHMM BED format files based on the hg19 genome to their corresponding positions in the hg38 genome using *liftOver*. This conversion requires a precomputed chain file mapping hg19 to hg38 coordinates, which can be downloaded from the UCSC Genome Browser (http://hgdownload.cse.ucsc.edu/goldenPath/hg19/liftOver/hg19ToHg38.over.chain.gz). Finally, we use *bedtools intersect* to perform the overlap analysis between the RNA transcription regions BED file and the chromHMM BED file.

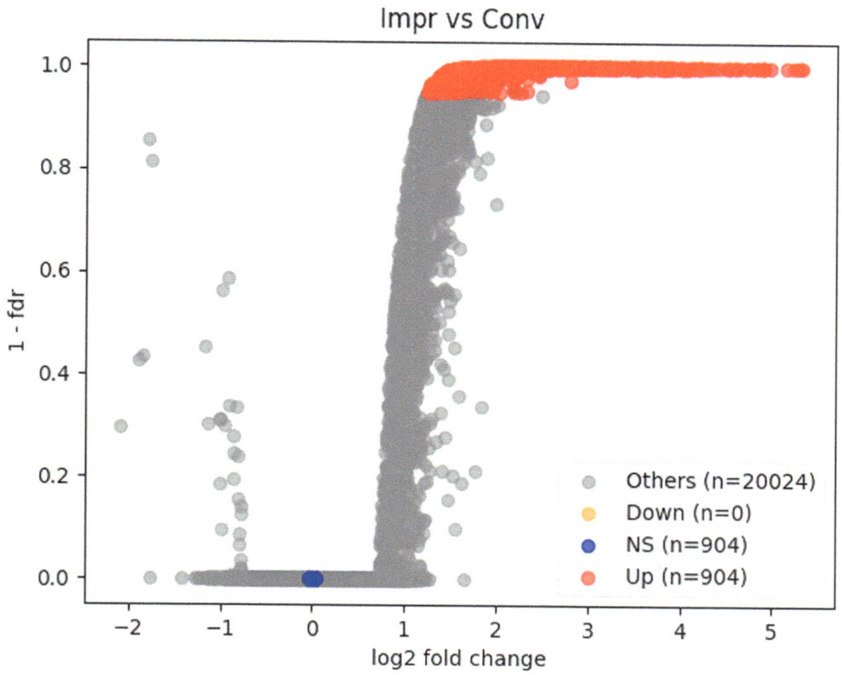

Fig. 1 Identification of semi-extractable RNAs. Impr, improved RNA-seq; Conv, conventional RNA-seq; Up, upregulated; Down, downregulated; NS, nonsignificant (a specified number of transcripts to match the number of upregulated transcripts, after sorting by |log2 fold change|)

```
$ liftOver chromHMM_hg19.bed hg19ToHg38.over.chain chromHMM_hg38.bed /dev/null
$ bedtools intersect -wo -a RNA.bed -b chromHMM_hg38.bed
```

By calculating the proportions of various chromatin states within the transcriptional regions of semi-extractable RNAs and comparing these values with those of extractable RNAs and all annotated (BG, background) RNAs, we observed that semi-extractable RNAs are more likely to originate from repressed (SE/EX = 3.53) and heterochromatin/repetitive (SE/EX = 2.74) regions (*see* Table 1).

3.4 Subcellular Distribution of Semi-extractable RNAs

We used *STAR* and *StringTie -e* to calculate the FPKM expression levels of each RNA in APEX-seq target and control samples across various subcellular fractions. The log2 ratio of these expression levels was then used to quantify RNA subcellular localization. Semi-extractable RNAs were observed to be enriched in nuclear compartments, including the nucleolus, nucleus, nuclear lamina, and nuclear pore (Fig. 2). Refer to *./bin/fig2.py* for plotting and p-value calculations.

Table 1
Preferential transcription of semi-extractable RNAs from repressed and heterochromatic regions

Chromatin states	%SE	%EX	%BG	SE/EX	SE/BG
Active promoter	0.88	7.24	4.38	0.12	0.20
Promoter flanking	0.45	1.66	1.13	0.27	0.40
Inactive promoter	0.07	0.24	0.23	0.27	0.29
Candidate strong enhancer	1.25	0.98	1.09	1.27	1.14
Candidate weak enhancer/DNase	4.05	4.05	4.05	1.00	1.00
Distal CTCF/candidate insulator	1.13	1.43	1.36	0.79	0.83
Transcription associated	12.70	40.69	27.13	0.31	0.47
Low activity proximal to active states	52.70	33.89	40.61	1.55	1.30
Polycomb repressed	3.25	0.92	2.72	3.53	1.20
Heterochromatin/repetitive/copy number variation	23.41	8.55	14.57	2.74	1.61

SE semi-extractable RNAs, *EX* extractable RNAs, *BG* background (total annotated) RNAs

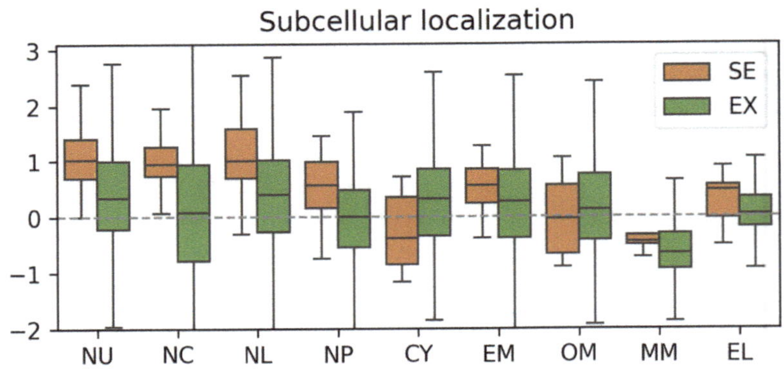

Fig. 2 Subcellular RNA Localization in HEK Cells Measured by APEX-seq Fold Changes. Higher values represent increased RNA abundance in specific subcellular fractions: SE (semi-extractable RNAs), EX (extractable RNAs), NU (nucleolus), NC (nucleus), NL (nuclear lamina), NP (nuclear pore), CY (cytosol), EM (endoplasmic reticulum membrane), OM (outer mitochondrial membrane), MM (mitochondrial matrix), and EL (ER lumen). Wilcoxon rank-sum test: $p < 0.001$ (NU, NC, NL, NP), $p < 0.05$ (CY), nonsignificant (EM, OM, MM, EL)

3.5 RNA–DNA Interactions Analysis of Semi-extractable RNAs

iMARGI is a high-throughput technique used to study RNA–DNA interactions within the cell nucleus. It identifies which RNA molecules are physically associated with specific genomic regions. The processed iMARGI results include genomic location information for interacting RNA and DNA. Using *bedtools intersect*, we can calculate the interaction frequency of each RNA with DNA, which reflects the strength of the RNA–DNA interactions. Refer to *./bin/fig3.py* for plotting and p-value calculations. Based on Fig. 3, we observed that semi-extractable RNAs exhibit weak RNA–DNA interactions.

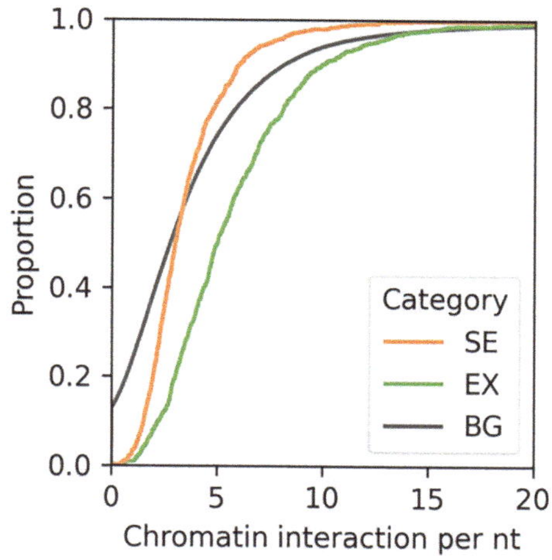

Fig. 3 RNA–DNA interactions measured by iMARGI in HEK cells. SE, semi-extractable RNAs; EX, extractable RNAs; BG, background (total annotated) RNAs. Wilcoxon rank-sum test: $p < 0.001$ (SE vs EX, SE vs BG, EX vs BG)

3.6 Minimum Free Energy (MFE), Length and GC-Content Analysis of Semi-extractable RNAs

We used MFE to quantify the stability of secondary structures at the ends of RNA sequences, which may affect RNA metabolism. Lower MFE values indicate more stable RNA secondary structures. We extracted 500 nt sequences from both ends of the RNA and used *RNAfold* to predict their MFE values. The MFE calculations, as well as the plotting and p-value computations, were performed in *./bin/fig4.py*. Additionally, comparisons and visualizations of RNA sequence length and GC content were implemented in the scripts *./bin/fig5.py* and *./bin/fig6.py*. It can be observed that the 3′ and 5′ ends of semi-extractable RNAs tend to form weak RNA secondary structures (Fig. 4). These RNAs also have longer sequence lengths and lower GC content (Figs. 5 and 6).

3.7 Density of Repetitive Sequences in Semi-extractable RNAs

We used the *bedtools intersect* tool to compare the regions of repetitive sequences annotated by *RepeatMasker* in the genome with the transcriptional regions of target RNAs. The script *./bin/fig7.py* calculated the coverage density of various repetitive sequences in each RNA. As observed in Fig. 7, semi-extractable RNAs are enriched with various types of repetitive sequences, including LTR, LINE, DNA, and SINE elements.

3.8 RNA–RNA Interactions Analysis of Semi-extractable RNAs

RIC-seq data provide insights into protein-mediated RNA–RNA interactions in the transcriptome, while PARIS data focus on direct RNA–RNA interactions. Based on these two datasets, we can identify lists of hub RNAs that interact with multiple RNAs. Using the script *./bin/fig8.py*, we visualized the overlap between hub RNAs and semi-extractable RNAs and calculated the corresponding significance levels, helping to determine whether semi-extractable

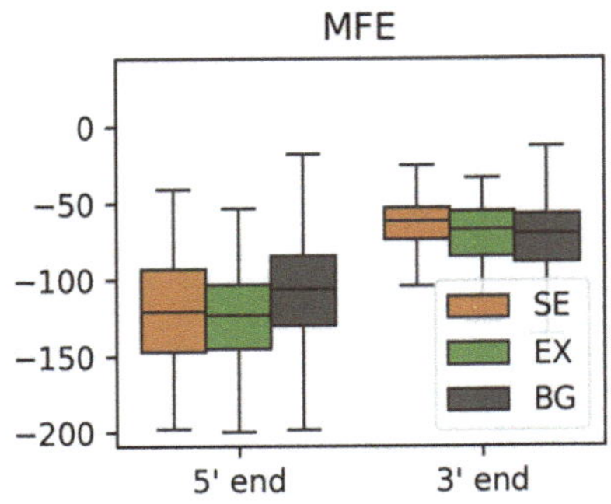

Fig. 4 Comparison of minimum free energy (MFE) in the 5′ and 3′ end regions. SE, semi-extractable RNAs; EX, extractable RNAs; BG, background (total annotated) RNAs. Wilcoxon rank-sum test: $p < 0.001$ (all SE vs EX, SE vs BG)

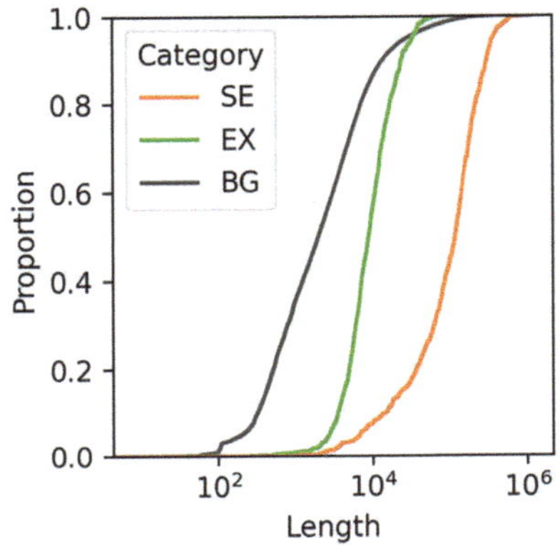

Fig. 5 Comparison of sequence length. SE, semi-extractable RNAs; EX, extractable RNAs; BG, background (total annotated) RNAs. Wilcoxon rank-sum test: $p < 0.001$ (SE vs EX, SE vs BG)

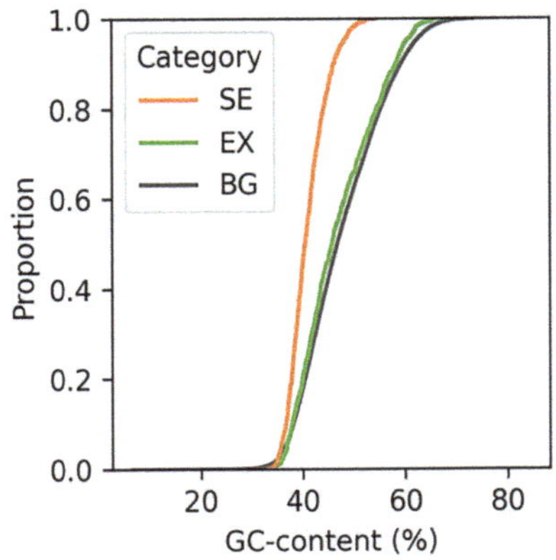

Fig. 6 Comparison of GC content. SE, semi-extractable RNAs; EX, extractable RNAs; BG, background (total annotated) RNAs. Wilcoxon rank-sum test: $p < 0.001$ (SE vs EX, SE vs BG)

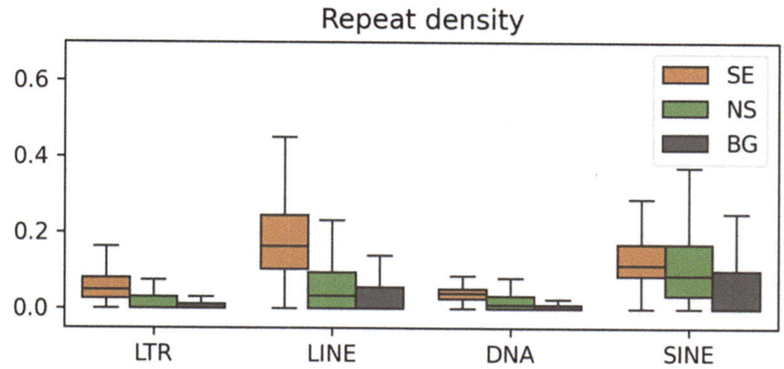

Fig. 7 Comparison of the coverage density of various repetitive sequences. SE, semi-extractable RNAs; EX, extractable RNAs; BG, background (total annotated) RNAs; SINE, short interspersed nuclear elements; DNA, DNA transposons; LINE, long interspersed nuclear elements; LTR, long terminal repeat. Wilcoxon rank-sum test: $p < 0.001$ (all SE vs EX, SE vs BG, EX vs BG)

RNAs are enriched in transcripts interacting with other RNAs. The list of hub RNAs precomputed from RIC-seq data can be accessed from the supplementary materials of the original paper. RNA–RNA interaction data curated from PARIS data can be obtained from the RISE database, where we further define transcripts interacting with 20 or more RNAs as hub RNAs.

```
$ awk '$21=="HEK293T" {print $12"\n"$14}' PARIS.txt | sort |
uniq -c | awk '$1 > 20 {print $2}' > PARIS_hubRNAs.txt
```

We observed that semi-extractable RNAs are significantly enriched with protein-mediated hub RNAs (Fig. 8a), whereas the enrichment of direct RNA–RNA interactions, as defined in the PARIS public dataset, was not observed (Fig. 8b).

3.9 RBP-Binding Profile of Semi-extractable RNAs

After dividing the RNA sequences into five equally sized bins, we used *FIMO* to scan for motifs of approximately 400 RNA-binding proteins (RBPs) recorded in CISBP-RNA. We then calculated the frequency of these motifs in the corresponding bins. Finally, we computed the log2 ratio of the motif frequencies between semi-extractable RNAs and extractable RNAs to determine the RBP-binding preference (*see* Table 2). The data processing and the generation of the table of RBP-binding preference were implemented using the script *./bin/table2.py*.

3.10 K-mer-Based Clustering of Semi-extractable RNAs

We used *seekr_kmer_counts* from the *seekr* package to calculate k-mer frequencies in semi-extractable RNAs, and *seekr_pearson* to compute a similarity matrix based on these k-mer frequencies.

```
$ seekr_kmer_counts SE.fa -o SE.freq -k 6
$ seekr_pearson SE.freq SE.freq -o SE.corr
```

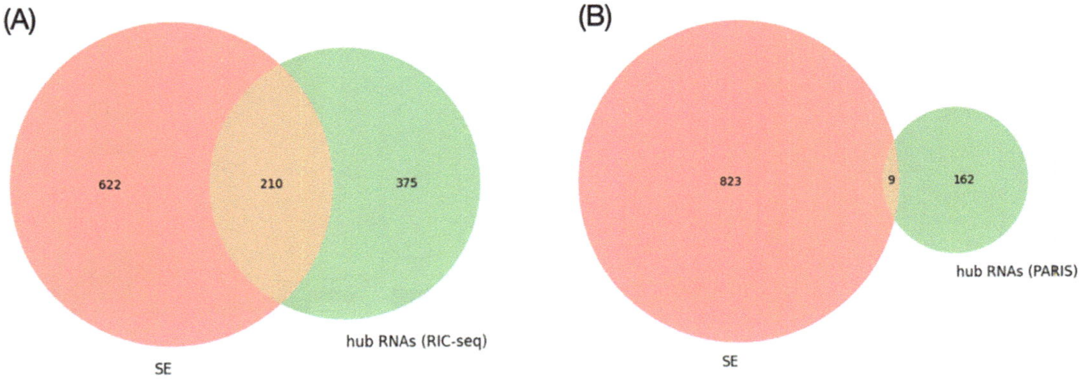

Fig. 8 Venn Diagram Analysis of Semi-extractable RNAs and hub RNAs of (A) RIC-seq and (B) PARIS. SE, extractable RNAs. Hypergeometric test: (A) $p < 2.30e{-}236$ (B) $p < 9.69e{-}04$

Table 2
Enrichment analysis of semi-extractable RNAs for RBP binding motifs across different positional regions, with extractable RNAs as controls

RBP	bin1	bin2	bin3	bin4	bin5	Average	
ELAVL2	1.07	0.692	0.585	0.439	0.114	0.564	Top 10
RBMS3	0.985	0.643	0.56	0.51	0.22	0.563	
A1CF	0.884	0.588	0.527	0.438	0.174	0.508	
HNRNPC	0.989	0.617	0.52	0.389	0.08	0.506	
HNRNPCL1	0.989	0.617	0.52	0.389	0.08	0.506	
RALY	0.98	0.628	0.524	0.387	0.072	0.506	
CPEB4	0.973	0.615	0.513	0.379	0.08	0.499	
RBMS1	0.827	0.594	0.482	0.449	0.204	0.498	
TIA1	0.946	0.596	0.488	0.367	0.061	0.481	
U2AF2	0.871	0.575	0.479	0.36	0.12	0.472	
FXR2	−0.676	−0.483	−0.407	−0.364	−0.127	−0.421	Bottom 10
SRSF4	−0.815	−0.56	−0.472	−0.441	−0.202	−0.514	
HNRNPH2	−0.933	−0.568	−0.49	−0.478	−0.288	−0.576	
SAMD4A	−0.838	−0.612	−0.565	−0.536	−0.305	−0.585	
SRSF7	−1.223	−0.822	−0.594	−0.504	−0.173	−0.707	
ZC3H10	−1.381	−0.943	−0.822	−0.741	−0.372	−0.901	
RBM84	−2.018	−1.398	−1.294	−1.174	−0.668	−1.493	
RBM4B	−2.076	−1.402	−1.159	−1.117	−0.759	−1.495	
RBM4	−2.203	−1.631	−1.389	−1.226	−0.722	−1.661	
PPRC1	−2.196	−1.612	−1.467	−1.364	−0.882	−1.711	

As the latest version of the seekr package does not provide functionality for clustering RNA sequences based on similarity matrices, we reimplemented this process in *./bin/table3.py*. We also recalculated the table of RBP-binding preferences for each RNA cluster. All results are saved in the *./output* directory.

4 Notes

1. In constructing the transcriptome, we exclusively utilized RNA-seq data from samples processed with the improved RNA extraction method. This approach minimizes the complexity of the merged transcriptome that can arise from including multiple samples. Furthermore, our analysis focuses on transcripts enriched by the improved RNA extraction method compared to the conventional method.
2. We generated a representative transcript for each gene by taking the union of the exonic regions based on the gene annotation file. To calculate the FPKM expression levels of each exon and intron, we temporarily treated these representative transcripts' exons and introns as transcripts, creating a GTF file. We then used *StringTie* and the mapping data to compute the expression levels. Subsequently, we calculated the RS value for each intron based on the mean expression of these exons and introns across all samples, determining whether the intron should be included in the representative transcript. The process of generating the GTF file is implemented in *./bin/collapse_gencode_gtf.py*, while the calculation of RS values is completed by *./bin/calculate_RS.py*.

Funding

AMED [JP22ama121055, JP21ae0121049 and JP21gm0010008 to M.H.]; JSPS KAKENHI [22H04925, 20H00624 and 17K20032 to M.H.]; and JSPS KAKENHI [20K15784, 22K15093 to C.Z.].

References

1. Chujo T, Yamazaki T, Kawaguchi T et al (2017) Unusual semi-extractability as a hallmark of nuclear body-associated architectural noncoding RNAs. EMBO J 36:1447–1462
2. Zeng C, Chujo T, Hirose T et al (2023) Landscape of semi-extractable RNAs across five human cell lines. Nucleic Acids Res 51(15): 7820–7831
3. McKinney W (2010) Data structures for statistical computing in Python. SciPy 445(1): 51–56
4. Harris CR, Millman KJ, van der Walt SJ et al (2020) Array programming with NumPy. Nature 585:357–362
5. Virtanen P, Gommers R, Oliphant TE et al (2020) SciPy 1.0: fundamental algorithms for

scientific computing in python. Nat Methods 17(3):261–272
6. Cock PJ, Antao T, Chang JT et al (2009) Biopython: freely available Python tools for computational molecular biology and bioinformatics. Bioinformatics 25(11):1422–1423
7. Hunter JD (2007) Matplotlib: a 2D graphics environment. Comput Sci Eng 9(03):90–95
8. Waskom ML (2021) Seaborn: statistical data visualization. J Open Source Softw 6(60):3021
9. Li S, Eberhard QE, Ni L, et al (2024) Improved functions for non-linear sequence comparison using SEEKR. RNA RNA-080188
10. Robinson MD, McCarthy DJ, Smyth GK (2010) edgeR: a Bioconductor package for differential expression analysis of digital gene expression data. Bioinformatics 26(1): 139–140
11. Li H, Handsaker B, Wysoker A et al (2009) The sequence alignment/map format and SAMtools. Bioinformatics 25(16):2078–2079
12. Dobin A, Davis CA, Schlesinger F et al (2013) STAR: ultrafast universal RNA-seq aligner. Bioinformatics 29(1):15–21
13. Kovaka S, Zimin AV, Pertea GM et al (2019) Transcriptome assembly from long-read RNA-seq alignments with StringTie2. Genome Biol 20:1–13
14. Liao Y, Smyth GK, Shi W (2014) featureCounts: an efficient general purpose program for assigning sequence reads to genomic features. Bioinformatics 30(7):923–930
15. Quinlan AR, Hall IM (2010) BEDTools: a flexible suite of utilities for comparing genomic features. Bioinformatics 26(6):841–842
16. Martin M (2011) Cutadapt removes adapter sequences from high-throughput sequencing reads. EMBnet J 17(1):10–12
17. Pertea G, Pertea M (2020) GFF utilities: GffRead and GffCompare. F1000Research 9
18. Lorenz R, Bernhart SH, Höner zu Siederdissen C et al (2011) ViennaRNA Package 2.0. Algorithm Mol Biol 6:1–14
19. Bailey TL, Johnson J, Grant CE et al (2015) The MEME suite. Nucleic Acids Res 43(W1): W39–W49
20. Ernst J, Kellis M (2012) ChromHMM: automating chromatin-state discovery and characterization. Nat Methods 9(3):215–216
21. Fazal FM, Han S, Parker KR et al (2019) Atlas of subcellular RNA localization revealed by APEX-Seq. Cell 178(2):473–490
22. Yan Z, Huang N, Wu W et al (2019) Genome-wide colocalization of RNA–DNA interactions and fusion RNA pairs. Proc Natl Acad Sci 116(8):3328–3337
23. Cai Z, Cao C, Ji L et al (2020) RIC-seq for global in situ profiling of RNA–RNA spatial interactions. Nature 582(7812):432–437
24. Lu Z, Zhang QC, Lee B et al (2016) RNA duplex map in living cells reveals higher-order transcriptome structure. Cell 165(5): 1267–1279
25. Gong J, Shao D, Xu K et al (2018) RISE: a database of RNA interactome from sequencing experiments. Nucleic Acids Res 46(D1): D194–D201

Index

A

B

C

D

E

F

G

H

I

L

M

N

O

P

R

Michiaki Hamada et al. (eds.), *RNA-RNA and RNA-DNA Interactions*, Methods in Molecular Biology, vol. 2949,
https://doi.org/10.1007/978-1-0716-4670-0,

MIX
Papier aus verantwortungsvollen Quellen
Paper from responsible sources
FSC® C105338

If you have any concerns about our products,
you can contact us on
ProductSafety@springernature.com

In case Publisher is established outside the EU,
the EU authorized representative is:
Springer Nature Customer Service Center GmbH
Europaplatz 3, 69115 Heidelberg, Germany

Printed by Libri Plureos GmbH
in Hamburg, Germany